Mom the Chemistry Professor

Renée Cole • Cecilia Marzabadi •
Gail Webster • Kimberly Woznack
Editors

Mom the Chemistry Professor

Personal Accounts and Advice from
Chemistry Professors who are Mothers

 Springer

Editors
Renée Cole
Department of Chemistry
The University of Iowa
Iowa City, Iowa
USA

Cecilia Marzabadi
Department of Chemistry & Biochemistry
Seton Hall University
South Orange, New Jersey
USA

Gail Webster
Department of Chemistry
Guilford College
Greensboro, North Carolina
USA

Kimberly Woznack
Department of Chemistry & Physics
California University of Pennsylvania
California, Pennsylvania
USA

Inspired by the German language book "Professorin und Mutter – wie geht das?" Published by Spektrum Akademischer Verlag, 2011.

ISBN 978-3-319-06043-9   ISBN 978-3-319-06044-6 (eBook)
DOI 10.1007/978-3-319-06044-6
Springer Cham Heidelberg New York Dordrecht London

Library of Congress Control Number: 2014942373

© Springer International Publishing Switzerland 2014
This work is subject to copyright. All rights are reserved by the Publisher, whether the whole or part of the material is concerned, specifically the rights of translation, reprinting, reuse of illustrations, recitation, broadcasting, reproduction on microfilms or in any other physical way, and transmission or information storage and retrieval, electronic adaptation, computer software, or by similar or dissimilar methodology now known or hereafter developed. Exempted from this legal reservation are brief excerpts in connection with reviews or scholarly analysis or material supplied specifically for the purpose of being entered and executed on a computer system, for exclusive use by the purchaser of the work. Duplication of this publication or parts thereof is permitted only under the provisions of the Copyright Law of the Publisher's location, in its current version, and permission for use must always be obtained from Springer. Permissions for use may be obtained through RightsLink at the Copyright Clearance Center. Violations are liable to prosecution under the respective Copyright Law.
The use of general descriptive names, registered names, trademarks, service marks, etc. in this publication does not imply, even in the absence of a specific statement, that such names are exempt from the relevant protective laws and regulations and therefore free for general use.
While the advice and information in this book are believed to be true and accurate at the date of publication, neither the authors nor the editors nor the publisher can accept any legal responsibility for any errors or omissions that may be made. The publisher makes no warranty, express or implied, with respect to the material contained herein.

Printed on acid-free paper

Springer is part of Springer Science+Business Media (www.springer.com)

Foreword

> "I have frequently been questioned, especially by women, of how I could reconcile family life with a scientific career. Well, it has not been easy."
> —Marie Curie, two-time Nobel Prize winner and mother of a daughter, Irène Joliot-Curie, who also won the Nobel Prize

While it may not have been easy being an outstanding scientist and mother, Marie Curie did both. The question of reconciling family life with a scientific career is still one that is asked today. Many women have combined successful careers as chemists in academia with motherhood; a few of them have shared their stories in this volume. While significant progress in increasing the number of professors who are women has been made, some significant challenges for all academic women chemists remain.

For example, a 2006 National Science Foundation (NSF) report indicated that female doctoral science and engineering faculty are less likely than their male colleagues (67 % vs. 84 %) to be married and also were less likely to have children living with them (42% vs. 50%) [1]. That report also indicates that the percentage of promotions to full professors of unmarried women and women without children from 1973 to 2006 was greater than the percentage of promotions to full professors of married women or women with children.

A 2010 article published in The Chronicle of Higher Education by Amy Kittelstrom, "The Academic-Motherhood Handicap," discusses the fact that academic mothers have different experiences than academic fathers, and that these differences are most significant during the intense years of childbearing and early caregiving—the years that matter most for academic careers [2]. Professor Mary Ann Mason from the University of California at Berkeley comments that if a woman wants to get hired as an assistant professor, she is much less likely to succeed if she is a mother. However, fathers are much more likely to land a position and achieve tenure, even more likely than childless men [3].

In the article "When Scientists Choose Motherhood," published in the American Scientist in March 2012 [4], one figure explains that the path to becoming a successful scientist looks much the same for women without children and for men with or without children: a straightforward, if long and arduous, track from undergraduate degree through postdoctoral and tenure-track positions. For women who have children, or even just plans to have them, the road is fraught with obstacles. Women who choose to have a first child are usually in the thick of the most difficult parts of their career. Even after the early, physically intensive months of childbirth and childrearing, women typically do more of the work of childcare and household management than do male partners, at the cost of productivity at work. As a result, they may choose to take industry or non-tenure-track jobs with less demanding hours, or they may leave the workforce entirely.

These reports tend to be the focus of most stories about combining academic careers and motherhood. Although the data and trends may paint a less than ideal picture for academic mothers, the Women Chemists Committee of the American Chemical Society (WCC ACS) enthusiastically welcomed the opportunity to provide inspiration, advocacy, and potential strategies for young women entering an academic career. We are very excited to be a part of this inspirational collection of stories from women chemistry professors with successful careers who are also mothers. This project fits very nicely into our committee's mission within the ACS.

The mission of the WCC is to be leaders in attracting, developing, promoting, and advocating for women in the chemical sciences in order to positively impact society and the profession. There are four goals designed to fulfill the mission statement:

1. Increase participation and retention of women in the chemical sciences and related disciplines
2. Take an advocacy position within the ACS on issues of importance to women in the chemical sciences
3. Provide leadership for career development opportunities for women in the chemical sciences and related disciplines
4. Promote and recognize the professional accomplishments of women in the chemical sciences and related disciplines

The WCC anticipates that this inspiring collection of heartfelt and real stories will increase the participation and retention of women who plan to become or who already are, "Mom the Chemistry Professor." While blending motherhood and professorship brings unique challenges, it is a rewarding path as evidenced by the stories in this volume.

References

1. Burrelli J (2008) Thirty-three years of women in S&E faculty positions: InfoBrief; NSF 08-308. National Science Foundation, National Center for Science and Engineering Statistics. http://www.nsf.gov/statistics/infbrief/nsf08308/nsf08308.pdf. Accessed 2 June 2014
2. Kittelstrom A (2010) The academic-motherhood handicap. The Chronicle of Higher Education: Manage Your Career. http://chronicle.com/article/The-Academic-Motherhood/64073/. Accessed 2 June 2014

3. Mason MA (2013) In the ivory tower: men only: for men, having children is a career advantage. For women, it's a career killer. Slate: Doublex. http://www.slate.com/articles/double_x/doublex/2013/06/female_academics_pay_a_heavy_baby_penalty.html. Accessed 2 June 2014
4. Williams WM, Ceci SJ (2012) When scientists choose motherhood. Am Sci 100: 138. http://www.americanscientist.org/issues/feature/2012/2/when-scientists-choose-motherhood/1. Accessed 2 June 2014

Madison, NJ, USA Amber Charlebois
2014

Contents

Part I Personal Accounts and Advice

Equilibrium and Stress: Balancing One Marriage, a "Two-Body Problem," and Three Children 3
Stacey Lowery Bretz

If at First You Don't Succeed, Don't Give Up on Your Dreams 19
Pamela Ann McElroy Brown

My Circus: Please Note That I Have No Formal Training in Juggling 29
Amber Flynn Charlebois

Planned Serendipity 43
Renée Cole

Mother and Community College Professor 53
Elizabeth Dorland

Chemistry in the Family 63
Cheryl B. Frech

Upward Bound to a Ph.D. in Chemistry and Beyond 73
Judith Iriarte-Gross

The Window of Opportunity 83
Nancy E. Levinger

Wanting It All 95
Cecilia H. Marzabadi

Taking an Unconventional Route? 103
Janet R. Morrow

From the Periodic Table to the Dinner Table 113
Danielle Tullman-Ercek

Encounters of the Positive Kind 129
Michelle M. Ward

The Long and Winding Road 139
Gail Hartmann Webster

I Finally Know What I Want to Be When I Grow Up 149
Catherine O. Welder

My Not-So-Secret Double-Life as a Chemistry Professor and Mom 165
Kimberly A. Woznack

**Remarkable, Delightful, Awesome: It Will Change Your Life,
Not Overnight but Over Time** 189
Sherry J. Yennello

Part II Safety in the Lab

Safety and Motherhood in the Chemistry Research Lab 201
Megan L. Grunert

Part I
Personal Accounts and Advice

Equilibrium and Stress: Balancing One Marriage, a "Two-Body Problem," and Three Children

Stacey Lowery Bretz

Introduction

As a young girl in the 1970s, I heard all about "superwomen." These were not women flying around in capes and tights, but they seemed just as powerful to me. Superwomen were so talented that they could have both a family *and* a career. No longer were young women expected to stay home with their children, nor forego children to pursue a career. It was now possible to 'have it all,' and I intended to do just that.

As with many childhood dreams, they are visions borne out of endless optimism and naïvete. I had no idea what an impossible standard I envisioned for myself. Nothing less than perfection! As a young girl, everything seemed possible, including having everything you ever wanted—having it 'all.'

I've since come to learn that chasing the elusive 'having it all' is folly and can create unrealistic pressures and disappointment for women and their families. No one can have it all. But, and this is a big BUT, you most definitely can have *both*. A family *and* a career. There is an important difference between having it all and having both. And that is the story I'd like to share with hopes that it inspires others.

S.L. Bretz (✉)
Department of Chemistry & Biochemistry, Miami University, Oxford, OH 45056, USA
e-mail: bretzsl@miamioh.edu

Middle School and High School (1980–1985)

I can still remember the first time someone told me I would not 'have it all'. My middle school biology teacher told me I wasn't strong enough to be a doctor. I would want to have children and choose to give up my career for staying home once children started to 'come along.' He made me so mad. Who was he to decide my future?! I remember thinking: I'll show him!

I did not end up in medicine, however, because in my sophomore year in high school, Mrs. Palmer introduced me to chemistry. She invited me to stay after school to join her team of students who conducted independent research projects. I had seen the 4 foot tall trophies these students won at the state research competitions. I doubted I could do original research as a 14-year-old. But Mrs. Palmer had confidence in me, and before I knew it, I had spent 3 years working in her classroom after school to explore the catalysis of urea and factors that affected the function of the enzyme urease.

That single invitation, and my teacher's unfailing belief in me, shaped my life in ways I could have never imagined. I won the 'Best Chemistry Project' award from Dow Chemical as a high school junior, which eventually led to a summer internship in college. My high school didn't offer AP courses (Advanced Placement), but conducting independent research made a strong impression and helped me get into Cornell University. But the most profound impact of that research experience was not professional—it was personal. There wasn't enough time after school to make all my solutions and then run my experiments. So, Mrs. Palmer invited one of her senior honors students (who had been on the state, trophy-winning teams) to make my solutions during his study hall so that I could get straight to work at the end of the school day. That boy was named Richard, and so began an incredible friendship that blossomed into young love. We were together for just 6 months before he graduated and left for college at Rensselaer Polytechnic Institute. Thirty-one years later, though, I am proud to say I am still married to my high school sweetheart and we've made a wonderful family. What follows in the rest of this chapter is the story of how we, as two chemists, have tried to balance our marriage, family, and careers.

College (1985–1989)

Stacey Lowery and Richard Bretz, 1983.

My first college course in chemistry was co-taught by Barbara Baird and Roald Hoffmann. Professor Baird was the only woman chemist on the faculty at Cornell at that time, and when the semester began, she was pregnant with twins. Professor Hoffmann is a Nobel Laureate. Looking back, I didn't realize then how exceptional my experience was. I just took it for granted then that of course you could be a woman professor and have a family, all the while on the tenure track. Of course other college freshmen were being taught chemistry by a Nobel Laureate. I was so naïve on both counts!

I went on to do undergraduate research in the lab of Héctor Abruña. All throughout college, I was still in a long-distance relationship with my high school sweetheart. And back in the days of no e-mail, no texting, and no social media—this meant writing letters. Lots and lots of letters. In my senior year at Cornell, Richard came to begin his PhD in the Abruña group. For awhile, we told no one we were dating (and had been for 6 years at that point!). We didn't know how that would be perceived. What if one of us would have to leave the group? After awhile, though, we felt comfortable letting the group, and eventually Dr. Abruña, know. (As I write this, I realize now that was probably the first, and only, time I ever 'hid' some part of my personal life from my colleagues.)

However, as I started looking toward graduate school, I learned that it was not OK to earn your PhD in the same department where you were an undergrad. It just wasn't done. So, after 1 year of togetherness, it was time to be apart again from Rich. I set off to earn a PhD at Penn State (3 hours away). The goal: to earn my PhD and become a faculty member at a small college where I could focus on teaching—and to do so in a location where Rich could either do the same or put his PhD in analytical chemistry to work in industry.

Grad School (1989–1994)

I had no idea how transformative grad school would be. I joined a research group that had never (ever) had a female graduate student. I had been warned not to do so by many people, but I was really interested in the chemistry. And more importantly, I was a confident woman. There was no way anyone or anything was going to derail my plans. I won Honorable Mention from the National Science Foundation (NSF) Graduate Research Fellowship program—clearly I had potential. I joined the group with two other guys, and they (and their projects) quickly outpaced me. I was assigned to an old lab by myself with a sometimes functioning piece of equipment. It took me a while to realize what was going on. Discrimination can be subtle. And denial can take a while to recognize and to admit the anger and the pain. Long story short, I dug deep. I got help with my instrument and data collection from the Abruña group back at Cornell and successfully defended my M.S. thesis before two committee members. My advisor chose not to come that day, saying he would pass me if my committee did. Which they did, and I graduated.

Which is not to say that the 2 years I spent at Penn State were nothing but frustration. To the contrary, some wonderful things happened while I was there and they set me on a new and exciting path that has shaped my career and my family. As a grad student, I was assigned to be a TA (teaching assistant) in first semester general chemistry. This was a very large course at Penn State (several thousand students), taught in lecture sections of 500 and small recitation sections of 25 or so. I loved being a TA. I was fascinated with the reasons why students had problems learning chemistry (and, to be blunt, why some professors could be so terrible at teaching it!). I wanted to investigate these problems and create solutions. I took a class on higher education from Mary Ellen Weimer and it opened my world to an entire literature I did not even know existed. So, I decided to make a change. Instead of looking for another PhD program in chemistry or another research group at Penn State, I wanted to get a PhD in chemistry education research (CER). Earning a PhD is the last opportunity for formal research training, and I wanted to learn how to investigate problems in teaching, learning, and assessment.

However, there were only M.S. programs in CER in chemistry departments then, so, I decided to return to Cornell. And then, something truly wonderful happened next. Because I was at Cornell again, Rich and I finally decided to marry—after 8½ years of dating. When Pastor Schieber welcomed our guests to our wedding, he started the ceremony by saying he knew what all who were present that day were thinking: "it's about time!"

Now, Cornell did not award PhDs in CER, but they did have a structure within the Graduate School where a student could create a novel interdisciplinary program with the endorsement of the faculty. At the time, my husband was a TA for Roald Hoffmann who suggested I contact Joseph Novak in the Department of Education at Cornell. Rich had taken a course with Joe and found it very interesting for thinking about a career in academia. I reached out to Joe, applied to Cornell, and soon had a committee of Joe, Roald, and Joän Egner who specialized in adult ed and higher ed. They helped me round out my chemistry coursework (Roald insisted I take biochemistry, which I am forever grateful he did. It fascinates me.) I took courses in curriculum & instruction, learning theory, and the history of science education in Education; survey design and program evaluation in Human Ecology; and applied statistics in Industrial and Labor Relations. I met colleagues who were embarking on careers in physics education research. One day at a seminar on physics education, I learned from a chemistry professor that Jerrold Meinwald, an organic chemist at Cornell, had just gotten a grant from the Mellon Foundation to develop a chemistry course for nonmajors and needed a research assistant (at this point I was filing papers in the Graduate School Admissions Office to pay the bills). I set off to schedule an appointment with Dr. Meinwald.

What I thought would be a short conversation with his secretary turned into a conversation to join him right then and there. An hour and a half later, he had offered me an RA (research assistantship) to help him design the course and was sending me to a brand new Gordon Conference on chemistry education to learn the latest and greatest goings on in the field. I had no idea how attending that one seminar would be such a pivotal point in my career...

I went to that Gordon Conference not even knowing what a Gordon Conference was. I had no idea how prestigious they were. I had no idea you must apply and be accepted to attend by the Chair. I had no idea I was the only grad student in the room. I had no idea who else was in the room. I didn't know any of these things. My naïvete was probably a good thing or I might have been too intimidated to say anything during discussion. I was simultaneously fascinated and frustrated by the first three talks. Really novel ways of teaching were being presented, but there were no data on how they impacted student learning. Most of the talks simply said "my students really seemed to like it." I raised my hand, noted that I found such claims to be odd because as chemists we would never accept liking something as a substitute for data, that there were most certainly methods for gathering and analyzing such data, and I hoped we would see some as the week went on. (Looking back, I cannot believe I had such chutzpah to do that! I was a new PhD student who thought she was learning the keys to the universe in her methods courses. But one could easily see how my enthusiasm might have been confused for hubris!).

Well, my comment (it never really was a question) turned out to be the last before we broke for lunch. I never got to stand up. I was surrounded by more people than I could count, all who were excitedly telling me about the planning grants they had just gotten from NSF for something called the Systemic Initiatives. (I told you I had no idea about who else was at this conference!). The NSF program officers were there, as were representatives from 14 different university collaborations. Each had $50 K planning grants to prepare the submission of their $5 M proposals to make radical changes in the undergraduate chemistry curriculum. They all needed to prepare detailed assessment and evaluation plans for these proposals. And more than a few were interested in what I knew about how to do such things. It was overwhelming to say the least.

The irony is that at a time when I was so very excited to be finding opportunities in this new discipline of CER, an unexpected event in my personal life happened that was completely beyond my control. A few weeks before the Gordon Conference, I found out that I was pregnant. This was not planned and definitely unexpected. We weren't sure how this was going to work with both of us still in grad school. But while I was out in California at the Gordon Conference, I started to miscarry. And even though I hadn't planned on getting pregnant, I was devastated to lose our first baby. I hadn't told friends the good news yet. How could I explain my near constant tears and sadness? It was a difficult time.

Postdoc (1994–1995)

I threw myself into my dissertation and the excitement of getting multiple postdoc offers to work on the Systemic Initiatives. Ultimately, I accepted an offer from the University of California, Berkeley. Rich and I graduated, packed up our two-bedroom apartment, and drove a rented moving van across the country.

Because the NSF had not yet officially made the award to Berkeley, I had to work on an alternative project for a few months. I was a postdoc with Angy Stacy in the Department of Chemistry, and we worked together on designing a new course for nonmajors. The grant was funded, and I became the Director of Evaluation, Assessment, and Pedagogy for a $5 M grant (as a postdoc!). I couldn't believe my good fortune to have secured such a position for the next 5 years.

Meanwhile, Rich was looking for employment, which everyone had assured us would be easy to find in the Bay area for a chemist. It took a long time. He taught part-time for a university and worked for an environmental company. In the give-and-take of a marriage, Rich was willing to support me in pursuing my 'dream job' first.

And while I learned a good deal of chemistry working with Angy, I learned even more from her example of how to combine motherhood with professorhood. When I walked into Angy's office on the first day, I remember my utter surprise at seeing a pack-and-go portable crib behind her desk. There were toys, the drawings of young children—things I had never seen before in a chemistry building, and certainly not in a professor's office. And to see this at a place with the stature of Berkeley! It forever changed my thoughts on what was possible. Angy didn't ask permission to bring her kids into the building. She just did it. We had many long conversations over the next year. I was especially keen to learn all I could about how she balanced family and career—for Rich and I had decided to have a child, and I was pregnant!

We thought we would be in California for several years, but in November a job ad appeared for a tenure-track position in CER at the University of Michigan-Dearborn. We knew that CER positions in chemistry departments were not commonplace. And Dearborn was only 2 hours from our hometown—and the four grandparents awaiting their first grandchild. We had to take a chance. I applied, interviewed when I was 6 months pregnant (which was obvious despite my carefully chosen suit), and was offered the job. Suzannah was born in May, and we drove back across the country in July.

Suzannah at age 11 months, surrounded by periodic tables tossed on the floor while her parents grade final exams at the kitchen table. She's holding both an HP and a Texas Instrument calculator in her hands.

Assistant Professor (1995–2000)

I began my tenure-track appointment with a 3-month-old child. My husband took an adjunct teaching position in the same department, and the chair agreed to not assign us courses at the same time so that one of us could always be with the baby. The plan was for Rich to continue teaching part time until Suzannah was old enough to go to the Child Development Center on campus (she had to be walking). I asked for and was assigned an office big enough to put a pack-and-go in. By February, I was pregnant with our second child who was due in late October.

When I've shared this story with young undergraduates or graduate students, I often get asked something akin to "weren't you worried they wouldn't think you a serious scientist to show up with a baby and get pregnant again during your first year?" And the simple answer is no. I wasn't worried. And, what's more, I didn't really care. My philosophy about family and work is pretty simple; I've never hidden my children, my wedding ring, pictures, toys, or kids' artwork. My children are an integral part of my life. If I have to hide them or not talk about them in order to be taken seriously at work, then that's not a place I want to work. And I have to say—that philosophy has served me very well throughout my entire career.

I was fortunate to work for a university with progressive policies. In addition to stopping my tenure-clock for 1 year, I was able to also benefit from a "modified duties" policy. This policy provides for circumstances when the 6–8 week recovery period *after* delivering a baby falls during an academic semester. This policy advises modifying one's duties from a combination of teaching/research/service to just research/service. So I didn't teach that fall. I worked on designing a new course instead. It was a fabulous policy and still exists at the University of Michigan. I have recommended it to many women at many universities in the years since.

Joey was born in October—nearly 3 weeks early. Rich continued to teach as an adjunct, and Suzannah went to the Child Development Center for a few mornings each week. By the time Joey was old enough to join Suzannah at the Child Development Center, Rich was itching to return to work full-time and was hired in a diesel engine research lab by Ford Motor Company. Because Rich had an "8–5" job, I was the parent with the more flexible schedule. I dropped off the kids and picked them up. My hours were not as long as everyone else's in the lab, but the moment they both were fast asleep, I went back into the office and worked many long nights for a few years. Life had reached a new 'normal' of sorts.

Joey at age 19 months reading a college chemistry textbook while Stacey prepares a lecture.

One time I returned from a conference, and a colleague asked which child had been sick for the last few days? I came to learn that because I wasn't in the office at the same hours as many of the other faculty, some had presumed I wasn't working or that I was home again with a sick child. I decided to 'announce' in the next faculty meeting that my schedule may not be typical, but that if colleagues needed to reach me, I returned to the building most every evening from 8 to midnight. To this day, whenever I travel professionally, I print and hang a sign on my door that says "Dr. Stacey Lowery Bretz will be out of town from mm/dd/yy to mm/dd/yy at University of XYZ or the ACS Conference or..." No one ever need wonder where I am again or default to the assumption that my absence must mean I am at home being a mother.

In early 1999, I was pregnant once more with what we hoped to be our third, and last, child, but once again, I miscarried. We really wanted a third child, but were learning once more we cannot dictate the timing of everything in our lives. I think this is a particularly challenging lesson for some academics to learn. We go to school. College is 4 years, grad school will be 5, postdoc for 2, then tenure clock for 5, etc., etc., etc. We have it all planned out and for the most part, we hit the milestones on time and just as we had planned them. Pregnancy and childbirth, however, are oblivious to our time clocks and planned milestones. And this can be a difficult thing to accept for most people, but I think especially for academics.

Associate Professor (2000–2005)

About that same time, I had realized that despite loving my job, I craved the opportunity to mentor graduate students in my research. I would go to conferences and see what colleagues were accomplishing with graduate students. I found myself with a severe case of "research envy." I needed to find a way to "move up the food chain" from an undergraduate institution to one that had a graduate program. This would be difficult to do given the expectations for research productivity and what I had been able to accomplish with limited resources at my current institution. But, in early 2000, I was invited to apply for a job opening at Youngstown State University. They were looking for someone to grow their M.S. program to include the discipline of CER.

When they called, I was already 8 months pregnant with our third child. I thought I would have to wait until after she was born, but that would be at least 2–3 months as I knew I would be having my third c-section. Ultimately, my obstetrician advised me to go now, *before* the baby was born. He wanted my husband to drive me there (4 hours away) just in case there were any unexpected surprises. But we also had a 4-year-old and a 2-year-old we couldn't leave behind. The solution? We stopped in our hometown on the way and picked up my in-laws. Yes, that's right. I went on a job interview 10 days before giving birth to our third child with my husband, my two young children, and my in-laws. We stayed in one suite at the Residence Inn. (I'll pause here while you stop chuckling.)

To say that I was 'huge' would be kind—it was my third pregnancy. (My first baby weighed over 9 pounds, and my son was over 11 pounds and nearly 23 in. long. This baby was going to be no different.) They broke a typical one-day interview schedule into two half-days. After just one meeting, they made everyone walk to the Chemistry Department, including the Deans, for fear that asking me to walk around campus that much would send me into labor (they only told me *that* more than a year later!). The job was great, and I wanted it. Just one problem: Rich was working at Ford and would have to give up a full-time job he loved to start job hunting once again in Youngstown, OH (not exactly a place of booming chemical industry). We talked it through, and once again, my husband put his career needs second so mine could take another important step forward. Mikaela was born in March 2000, and we moved during the summer.

Mikaela at age 5 dressed up in goggles, gloves, and a lab coat, to help Stacey with a liquid nitrogen experiment for her kindergarten class.

Rich once again took up part-time teaching in the chemistry department and began to volunteer with ACS Career Services. He found it very fulfilling and was becoming an expert on the "two-body" problem. One challenge we faced repeatedly was how to travel to conferences. Childcare was nonexistent at ACS meetings at that time. We tried lots of different "arrangements," but none was a permanent solution. We brought grandparents with us. We left kids at home with grandparents. We 'divided up' the meeting—he went for the first few days, I went for the last few days, and we literally were up in the air in two different planes at the same time—hoping for one of us to make it home in time to meet the school bus. We hired an undergraduate to join us and paid her travel and a stipend to watch the kids and entertain them while we were both busy at the same time. One of us went, and the other stayed home, missing the entire meeting. And, sometimes we just took the kids with us, no help whatsoever, and we juggled at the meeting in real time. My kids learned early on how to navigate the expositions at ACS meetings and how to sit quietly in the back of a research talk that Momma was giving. They've been to San Diego, San Francisco, Anaheim, Denver, Salt Lake City, Washington DC, Orlando, Boston, New York, Philadelphia, Chicago, and probably other places I'm forgetting. Did I miss out on some great talks? Yes. Did I miss opportunities to go out for drinks with colleagues and form new collaborations? Probably. But, did our family get to visit zoos and museums and try new restaurants, and see how many forms of transportation we could take on one trip? Yes! And we've made great, great memories doing so. (Now as teenagers, they beg to stay home and 'take care of themselves,' urging Rich and me to go by ourselves and 'have fun!')

Equilibrium and Stress: Balancing One Marriage, a "Two-Body Problem,... 13

From *left* to *right*: Stacey, Michelle (an undergraduate student who babysat for the family), Mikaela, Suzannah, Rich, and Joey getting ready to "Ride the Ducks" in Philadelphia during an ACS meeting in 2004.

Rich was soon hired in the Provost's Office to work on initiatives for the recruitment and retention of underrepresented students. He was instrumental in the opening of the first Early College High School in Ohio on the Youngstown State Campus, an initiative funded by the Bill and Melinda Gates Foundation. We decided to add to the balancing act by bringing two basset hound puppies into our lives—Rensselaer and Cornell. We had once again established a new equilibrium.

Which is not to say there weren't unexpected 'hiccups' in the daily routine. Children get sick and can't go to school. Or, they get pink eye and aren't permitted in school. More than once I worked from home on days I wasn't teaching. Or, I brought a child with me to sit in the front of my classroom while I taught class. They loved to bring their backpack of crayons and coloring books and sit in a desk "like the big kids." One time, Suzannah toddled up to the board to answer a question about equilibrium that had stumped the class for a few seconds!

Professor (2005–??)

By 2005, my career was progressing rapidly. I was elected by my peers to chair the Gordon Research Conference on Chemistry Education Research and Practice—the very same GRC I had attended just 12 years earlier as a bright-eyed young grad student. (It still seems surreal to me sometimes...)

I was also offered the opportunity to move to Miami University and have PhD students in CER. There were fewer than two dozen chemistry departments in the country where I could do this. To have a job offer from the one program in Ohio was tempting. But accepting this job would once again mean asking Rich to step down from a full-time job he found exceptionally fulfilling both personally and professionally. Miami and Youngstown were 5 hours apart. Commuting was not an option. Living apart was not, either.

We both went back to Miami on the second visit trip after they made me the offer. Not only were we going to negotiate my position, but we were going to negotiate one for him, too. Once again, we took our children and that meant bringing my in-laws to watch them while we were negotiating (I combine my career and motherhood in lots of ways, but trying to negotiate space and salary while wiping a nose or changing a diaper seemed a bit much, even for me!) I put to use every strategy I had learned in the COACh workshops (Committee for Advancement of Women Chemists). They worked, and worked well. (When the meeting was over, my department chair asked me where I had learned to do that. He then wanted to know if COACh would ever let men attend their workshops!) In the end, we chose to move to Miami.

Rich was hired as a visiting assistant professor. It was full-time work, but only temporary. The job hunt for a permanent position for my husband was on once again. During the next 3 years, he moved into a half-time administration/half-time instructor position, but in reality, it was like he had two full-time jobs.

When we moved to Miami, our youngest was just starting kindergarten. Never once have all three children been in the same school building and on the same schedule in terms of when they get on the bus and when they get off the bus. We've "split" the day—Rich is an early riser, so he would get up and head into the office early. I'd sleep a bit later and then drop off the children at the elementary school (we transferred our children from the elementary out near our township home to the one closest to the university). He would pick them up and bring them back to the department. We convinced the department chair to move Rich's office next to mine. The kids would play on our computers, eat snacks from our always stocked mini-refrigerators, and wait for us to decide it was time to go home. Eventually, we were able to convince the school district to route one of the buses through campus. Our middle-school children enjoyed the independence of walking one block to our offices from the bus stop.

Richard Bretz, Assistant Professor of Chemistry, Miami University, Hamilton, OH.

I am very proud to share that in 2012, Rich was hired as an Assistant Professor of Chemistry at the Miami University Hamilton campus. Once again he is working with nontraditional students and our family is "back on the tenure-track." It's my turn to make sure if there are any sacrifices to be made in terms of schedule or family conflicts that his career takes priority now.

Moving to Miami catalyzed my research into high gear. The CER program at Miami has become one of the top in the country. I have an amazingly talented colleague named Ellen Yezierski who makes going to work each day exciting and challenging. I've learned a lot from he, and she has become one of my closest friends. I've had the privilege to mentor three postdocs, 12 graduate students, and dozens of undergraduate research students. These talented people have helped amass a record of 60 publications, over 120 invited seminars, nearly 200 conference presentations and posters, and research findings that are changing what we know about how to measure student learning. I also know they've watched and learned from Rich and me that balancing motherhood with a career as a professor is an ongoing dynamic equilibrium. Some days are more in balance than others. Our family dynamic has changed many times over the years, with each age and school year of our children bringing new challenges. This year brought a brand new challenge: our oldest Suzannah is now a freshman at Cornell. How to get a child moved into college when both parents need to be at their own universities for the first week of the school year? Just when we think we've figured it all out, we learn we have more to learn...

Having it All vs. Having Both

Suzannah, Mikaela, and Joe, Christmas 2013.

I started this chapter by evoking the memory of the 1970s and 1980s "superwomen" who could "have it all"—family *and* a career. While I used to think that possible, what I've learned is it's an unrealistic standard to try to live up to. To put pressure on oneself to never fall short at work and to never fall short at home—that's a recipe for stress to be sure. You will miss an important conference or talk. You will miss the occasional spelling bee or field trip.

But, you can have *both* a family and a career. And you can find both to be incredibly fulfilling. You will choose to miss a faculty meeting so you can cuddle your sick child on the couch. You will explain to your children that you cannot stay home with them just because school was canceled due to a snowstorm. The University is still open and you must teach your class because that's how Momma and Daddy pay the bills.

In the end, it is pointless to worry about what your colleagues think of your choices. The only people you need answer to are you and your family. As Dr. Abruña told us, there is no such thing as a perfect time to start a family. If you wait for that perfect time, you may well end up childless. I've learned that children grow up very quickly. There will be plenty of time to spend grading papers and going to conferences once they're out of the house. Then it can be all the work I want. In the meantime, I plan to continue indulging in both my family and my career. The two are inextricably linked, and both bring me great joy.

Right before we moved in the summer of 2000, the Biennial Conference on Chemical Education was held at the University of Michigan. One night there was a social event held at the Henry Ford Museum and Greenfield Village in Dearborn. I went to the reception and took Mikaela (who was just 4 months old) with me. My colleague, and dear friend, Marcy Towns was also at that meeting with her son, Jimmy, who was just 6 weeks older than Mikaela. At one point in the evening, we both wandered off into a remote corner of the museum, looking to find a place where we could quietly breast-feed our babies. Neither one of us expected what happened next. Slowly, one by one, women graduate students and young women faculty approached us, often timidly asking "Can I ask you a question?" They had many questions on their minds. When did you tell your chair you were pregnant? How did you decide when to have a child—pre tenure or to wait until after? How did you avoid being in the teaching lab if there were chemicals thought to be harmful to your baby? Before long, Marcy and I realized we were surrounded by nearly 20 young women looking to us for sage advice on combining motherhood with being a professor. We certainly didn't pretend to have all the answers, or even any answers that might be useful in the circumstances of any one of those young women. All we could do is tell our story, as I do here once again. In doing so, it is my hope that some of you have found a new thought or idea here that might be useful to you and your family in the search for balance and equilibrium.

Rich and Stacey at her 25th reunion for the class of 1985 from Perkins High School, Sandusky, OH.

Main Steps in Stacey's Career

Education and Professional Career

1989	B.A. Chemistry, Cornell University, NY
1991	M.S. Chemistry, Penn State University, PA
1994	Ph.D. Chemistry Education Research, Cornell University, NY
1994–1995	Postdoctoral Fellow, University of California—Berkeley, CA

1995–2000   Assistant Professor, University of Michigan—Dearborn, MI
2000–2005   Associate Professor, Youngstown State University, OH
2005–present   Professor, Miami University, OH

Honors & Awards (selected)

2009 –present Chair of the Board of Trustees for the American Chemical Society Division of Chemical Education Examinations Institute
2010   American Association for Advancement of Science, Fellow
2012   American Chemical Society Fellow

Stacey has dedicated her professional life to teaching chemistry and researching chemistry education. She has been a member of the National Academy of Science National Research Council's Committee on Discipline-Based Education Research.

If at First You Don't Succeed, Don't Give Up on Your Dreams ...

Pamela Ann McElroy Brown

In the Beginning

I first fell in love with chemistry in high school, admiring the patterns in the periodic table which allowed prediction of the properties of the different elements, enjoying the challenge of balancing equations and stoichiometry, and recognizing the potential for solving societal problems with chemistry. Even before that I had fallen in love with babies. My sister was born when I was 10 years old and we shared a room. Her sweet smile and joy at discovering the world were irresistible.

My goals when I left for college were to get a degree, maybe even a master's, get married, work for a few years, have a baby, stop working, and in the 5 years before kindergarten earn a PhD. I would then get a job as a college professor when my child started school, where I could continue to conduct research and help to educate the next generation. I really wanted to be around in those important preschool years which I had seen with my sister were precious but fleeting. It did not matter to me that no woman in my family had even attended college, this was my plan! My own mother, a farm girl from central California, had dropped out of high school after 10th grade and was married on her 17th birthday, the first day she was legally able to do so without parental permission. I was born when she was still a teen and was often mistaken for her younger sister. She felt trapped by her lack of education and

---

P.A.M. Brown (✉)
New York City College of Technology – City University of New York, 300 Jay Street, Brooklyn, NY 11201, USA
e-mail: pbrown@citytech.cuny.edu

domestic responsibilities, and constantly urged me to get an education so that I could be economically self-sufficient. My father was 8 years older than her and earned a good living as a radio announcer. He felt, however, that since he was the sole wage earner, all childcare and housework were my mother's responsibility. Finding a husband when I was in college, who would help with the housework and childcare and was supportive of my dreams, was an important component of "the plan."

The Plan Unfolds

I began dating my future husband Harvey in college. After earning my BS in Chemistry from the State University of New York at Albany, my tuition and living expenses were paid for by a Gulf Oil fellowship while I worked toward a master's degree at the Massachusetts Institute of Technology (MIT). After finishing my master's I was employed a year at American Cyanamid doing research on hydrodesulfurization catalysts. We saved enough to get married and buy our first home. I needed to relocate as my husband worked in the family demolition business and did not have that flexibility. I was able to find employment at Pall Corporation, a manufacturer of medical and industrial filters, working in technical service. This was my dream job. I would arrange for potential customers to send samples to our labs to determine which of the products would best meet their filtration needs. I would also travel to their facilities and deliver equipment so that they could evaluate the filters themselves under process conditions. I would help existing customers if their specifications changed or if they encountered problems. Once I discovered I was pregnant I decided that I would return to work after a few months of maternity leave. I enjoyed my job, had family living nearby to help out, and also felt the financial pressures of home ownership.

Motherhood

My daughter Heather was born when I was 26. She was beautiful beyond words and seemed perfect. Unfortunately when she was a few weeks old it was discovered that she had a congenital birth defect—a dislocated hip. She was born without a hip socket and if left untreated would walk with a limp and develop painful arthritis at an early age. The treatment was a brace that needed to be worn essentially 24 h a day, 7 days a week for a year or more. The brace pushed the legs into the pelvis, slowly creating a hip socket in the soft, infant bones. If the brace wasn't readjusted correctly after diaper changes, etc., the treatment would be ineffective. I decided to resign from my job and stay home and take care of her. My second daughter Vanessa was born 17 months later. When the girls were 18 and 35 months old I returned to college to begin work on a PhD. I had been awarded a Teaching

Assistantship which paid for my tuition and provided a stipend for expenses, which I used for childcare. I took classes in the evenings and taught a 6 h lab on Fridays. Because I already had a master's degree and had also taken graduate courses as an undergraduate, I was able to complete my required coursework in one academic year and pass the qualifying exams. I made arrangements with one of the faculty members to begin research on my dissertation the following fall, again receiving a stipend and tuition paid through a research assistantship. This is one of the many advantages of pursuing graduate work in the chemical sciences—the opportunities to earn a degree with no out-of-pocket expenses in exchange for working as a teaching or research assistant. My education was supported by the National Science Foundation and a James Lago Fellowship.

My dissertation project involved a novel process for the purification of terephthalic acid (TPA), which is used to synthesize the long chain polyester, polyethylene terephthalate (PET). PET is used to make products ranging from plastic soda bottles to fabric. The PET produced today is much stronger because it is made from purer TPA. My work contributed to this improvement.

A little over 3 years after starting research I successfully defended my dissertation. During this time I became an efficiency expert—I would plan meals a week in advance, so I would only need to make one trip to the supermarket. While my fellow graduate students would socialize in the mornings, I would get right to work, arriving early and leaving late to avoid rush hour traffic. About a year before finishing my dissertation my son William was born. Fortunately, all experimental work had been completed and my advisor, who was very supportive, allowed me to finish up from home and continued my financial support, although at a reduced level. During naps I finished writing computer simulations and my dissertation. After earning my PhD I was exhausted and took several months off. I then continued to work for my advisor for a couple of years on a part-time basis as a consultant—I would come in and run experiments, do literature surveys, and write reports. I also published two papers on my dissertation research in peer-reviewed journals during this time.

There were many people along the way who served as my mentors. As an undergraduate at the State University of New York my Physical Chemistry professor, Dr. A.J. Yencha, provided me with the opportunity to conduct research with him. He provided invaluable advice when I applied to graduate school. My PhD dissertation advisor, Dr. Alan Myerson, was extremely understanding and supportive of my family needs. He helped me to develop a research project where much of the work could be done at home using the computer. He made sure that I had access to the resources needed in the laboratory. He was a brilliant scientist combining his responsibilities as department chair with an active research program. He was patient when experiments led to a dead end and helped me to develop more successful approaches.

A Career in Academia

I decided that a career in academia would allow for good work–life balance and was a better fit for my interests and the realities of the local economy. The chemical industry in the metropolitan New York area was shrinking and positions were few and far between. The academic calendar was much more flexible than that of the corporate world. When my son was in kindergarten I started a one day a week position teaching a physical chemistry laboratory at Barnard College. After a year I was offered a full-time position as a visiting assistant professor at Stevens Institute of Technology. It was a 9 month a year non-tenure-track teaching appointment. I hadn't even applied for the position but had been highly recommended by my dissertation advisor when someone else left just before the start of the semester. I was thrilled! My coworkers were nice and the students were a delight to work with. I stayed at Stevens for 4 ½ years but became increasingly frustrated by the lack of job security and professional growth potential—there was always uncertainty waiting for another annual reappointment, and no opportunities to conduct research. My salary was also considerably less than that of the tenure-line faculty. I enjoyed developing new curricular materials and managed to present at educational conferences and publish in educational journals. One of my happiest memories was taking two of my children with me to present at a conference in Washington, DC.

Finally, I was offered a tenure-track position as an assistant professor at New York City College of Technology (then New York City Technical College), a branch of the City University of New York, in a 2-year chemical technology program. The department, Physical and Biological Sciences, included faculty in biology, chemistry, and physics. By now my children were 16, 14, and 10—they were growing up fast! Since my commute was 90 minutes each way, arranging car pools to after school activities was a perennial challenge. Meals were planned a week in advance and fortunately my oldest daughter enjoyed cooking and frequently prepared dinner. When she was busy we often had "frozen food feasts" to save time but still have meals together as a family. After dinner everyone would finish their homework at the kitchen table and I would help them out as needed. Their academic success was a high priority. Harvey was also a big help despite his long hours. I continued to have childcare available after school and on school holidays, but my oldest daughter in particular was beginning to resent having a "stranger" in the house.

There are three major areas of responsibility for tenure-line faculty—teaching; service to the department, college, and university; and scholarship. In 2-year programs the major focus is on teaching and service, although there is the expectation of scholarship as evidenced by publications, presentations, and other scholarly work. Faculty employed at the City University of New York, which includes community colleges, 4-year institutions, and graduate and professional programs are employed under a collective bargaining agreement. The expectations of productivity in all three areas are the same, although the emphasis is different depending on the focus of the institution. When I began, my teaching load was

27 h annually, typically 15 h one semester (five sections meeting 3 h each week) and 12 the other (four sections meeting 3 h each week)—teaching and service were the main focus. I also advised students, helped those in my classes during office hours, was the faculty advisor to the chemistry club, wrote letters of recommendation, helped students obtain internships, and served on several campus-wide committees. I ran for College Council, the faculty governance body, and eventually chaired the committee making policies related to students. I began a research program with undergraduates, presented at regional and national conferences, and continued to develop new curricular materials and publish in educational journals. Every summer I applied for, and was awarded, grants through the American Chemical Society's Project SEED, which provided generous stipends to promising but underprivileged high school students. The undergraduates developed their own mentoring skills working with the SEED students and had fun. One project even resulted in a publication in the *Journal of Undergraduate Chemistry Research* while one of the contributors was still in high school!

After 3 years I applied for and was awarded a promotion to associate professor and after 5 years I had tenure at long last! My two daughters had graduated high school and went off to college and my son was now in high school—where had the years gone?

Teaching in a 2-year program allowed me the opportunity to teach the subject that I loved and really make a difference in the lives of my students. Many of the students that I worked with went on to graduate and professional schools. Some are even adjunct chemistry faculty members on my campus—sources of inspiration to current students. Mentoring promising high school students was a delight. A moment I will never forget is when I heard a knock on my office door one summer while on the phone. Not wanting to get up I asked, "Who is it?" My SEED student responded, "A budding scientist."

When the Chemical Technology Program Coordinator was appointed as Acting Dean, I became the Program Coordinator. In addition to teaching, my responsibilities now included scheduling chemistry courses and faculty, interviewing and recommending adjunct faculty for part-time teaching positions, handling student complaints, evaluating transfer credits, and facilitating monthly meetings with all chemistry faculty members to discuss departmental issues such as curriculum development, text book selection, and social activities. I set up the annual meeting with our External Advisory Board, made up of alumni and other representatives of local industries, to assure that our curriculum met workforce needs. I continued to conduct research with undergraduates and involved high school SEED students during the summer.

Sixteen months after being asked to be the Program Coordinator I was asked to serve as Acting Dean of the School of Arts and Sciences. Two years later after a national search I was appointed dean. In that capacity I oversaw eight academic departments. Part of my responsibilities included fund raising through grant writing. I was the principal investigator on a $1 million National Science Foundation (NSF) grant to increase the number of STEM graduates and was the co-PI on several other NSF grants, totaling over another $2 million.

After 6 years as dean, I had the privilege of serving as a Program Director for the National Science Foundation in the Division of Undergraduate Education, in Arlington, Virginia, under a 1-year appointment. The National Science Foundation is the largest government agency responsible for supporting external research. Scientists and engineers submit proposals which are reviewed by experts in the field, to help assure that only the most meritorious are funded. In my capacity as a Program Director I managed a portfolio of ~50 current awards and made funding recommendations totaling ~$13.5 million. I identified reviewers, assembled and oversaw review panels, and contributed to new solicitations. I also traveled around the country providing outreach to the larger undergraduate education community through presentations on funding opportunities, NSF initiatives and also offered grant writing workshops.

Upon my return to the college I was promoted to associate provost. My responsibilities now include oversight of program review and reaccreditation, program and curriculum development, faculty professional development, and coordinated undergraduate education.

My oldest daughter, Heather, has chosen a life in academia and is a lecturer in Health Economics at Newcastle University, in Newcastle, England. She earned a PhD from Sheffield University in England and is married to Adam. They met in Scotland while she was doing a postdoc. They recently became parents to baby Sadie. Vanessa works in Human Resources and lives in Manhattan, New York. William graduated from law school and works in labor law on Long Island, New York. My husband Harvey and I are still happily married, enjoy traveling and local cultural activities, and are looking ahead to retirement in a few years.

While I consider raising my children to be my greatest life work, my career was always a source of great pride and joy. A job in academia allowed me to do both. My oldest daughter tells me that my insights helped her earn her PhD and obtain her position. She teaches graduate students and her research is her most important professional priority. Vanessa fondly remembers Saturdays at science fairs and William always enjoyed the kitchen table science experiments I tested on him before demonstrating to my classes (red cabbage juice is a great pH indicator!).

Throughout my professional life, mentors continued to play an important role. As mentioned earlier, once I became the Dean of Arts and Science I wrote a successful National Science Foundation $1 million STEP grant to increase the number of graduates in science, technology, engineering, and mathematics (STEM) at my institution. My Program Director, Dr. Susan Hixson, provided much needed advice on how to successfully manage a large, multidimensional grant. Through annual meetings for principal investigators, I became part of a national community dedicated to student success, sharing information and ideas. I was asked to review proposals and met Program Director Elizabeth Dorland, who purposely arranged connections within the chemical education community. Through her intentional networking I met Dr. Rick Moog, principal investigator (PI) on several NSF grants and was able to introduce Process Oriented Guided Inquiry Learning (POGIL) on my campus. I also met Dr. David Burns, another PI, and helped my campus join the Science Education for New Civic Engagements and

Responsibilities (SENCER) community. Both initiatives led to innovative curriculum and improved teaching.

The greatest challenge that I encountered in my career was obtaining a tenure-track position once I returned to the workforce full-time. While tenure-line positions were once the norm, there is a growing shift towards part-time instructors. While some are employed full-time in industry and teach part-time for personal satisfaction or supplemental income, many patch together full-time employment by working part-time at multiple campuses. For those in this situation, there is little security, much hard work, few if any opportunities for research and scholarly growth, and typically annual salaries much lower than those of tenure-line faculty. At my campus about 50 % of the classes are taught by part-time faculty members.

In my opinion the best way to prepare for a tenure-line position is to speak to as many people in academia as possible to learn about the culture and expectations. Ask about the hiring process and the characteristics of successful applicants. Peer-reviewed publications are a benchmark of scholarly accomplishments. Plan for publishing even before commencing research. Read the journals and model your writing after successful authors. Obtain teaching experience—you can start as a tutor or peer mentor as an undergraduate and serve as a teaching assistant (TA). Look for teaching opportunities in graduate school. Read journals such as the *Chronicle of Higher Education* to learn more about the issues facing higher education today. Assessment of student learning and cycles of continuous improvement are mandates of the accrediting agencies. Learn about assessment and if possible get solid experience. Read journals such as the *Journal of Chemical Education* to learn more about pedagogy, curriculum development, and effective practices in the class room. Learn about strategies for writing proposals—in many cases grants will fund your research and other projects. In summary, academia is a challenging career, with many demands, requiring years of preparation. It is also extremely rewarding, with opportunities for work–life balance and personal satisfaction.

Interview with the Author

1. How has deciding to start a family or having a family influenced your career? How has your career influenced your family?
I decided that a career in academia would allow for good work–life balance and was a better fit for my interests and the realities of the local economy.

2. Did you have role models? Which examples were set for you in your childhood or while you were growing up?
There were many people along the way who served as my mentors. As an undergraduate at the State University of New York my Physical Chemistry professor, Dr. A.J. Yencha, provided me with the opportunity to conduct research with him. He provided invaluable advice when I applied to graduate school. My PhD dissertation advisor, Dr. Alan Myerson, was extremely understanding and supportive of my family needs. He helped me to develop a research project

(continued)

where much of the work could be done at home using the computer. He made sure that I had access to the resources needed in the laboratory.

3. Have you come up against any significant obstacles during your career and how did you overcome these?

The greatest challenge that I encountered in my career was obtaining a tenure-track position once I returned to the workforce full-time. While tenure-line positions were once the norm, there is a growing shift towards part-time instructors. I accepted non-tenure-line positions to gain experience and kept trying.

4. Is there anything you would have done differently or would not do again?

I always try to look forward.

5. What advice would you give to young women hoping to pursue a career in academia? E.g., while studying, when planning a family

In my opinion the best way to prepare for a tenure-line position is to speak to as many people in academia as possible to learn about the culture and expectations. Ask about the hiring process and the characteristics of successful applicants. Peer-reviewed publications are a benchmark of scholarly accomplishments. Plan for publishing even before commencing your research. Read the journals and model your writing after successful authors. Obtain teaching experience—you can start as a tutor or peer mentor as an undergraduate and serve as a TA. Look for teaching opportunities in graduate school. Read journals such as the *Chronicle of Higher Education* to learn more about the issues facing higher education today. Assessment of student learning and cycles of continuous improvement are mandates of the accrediting agencies. Learn about assessment and if possible get solid experience. Read journals such as the *Journal of Chemical Education* to learn more about pedagogy, curriculum development, and effective practices in the classroom. Learn about strategies for writing proposals—in many cases grants will fund your research and other projects.

About the Author

Pamela Brown is currently the Associate Provost at New York City College of Technology, a comprehensive college offering both associate and baccalaureate degrees, and one of the 23 branches of the City University of New York. Prior to this Pamela was the Dean of the School of Arts and Sciences, after having served as the Program Coordinator for Chemical Technology, a 2-year degree program. She continues to hold the rank of associate professor in the Chemistry Department. In 2011–2012 she had the privilege of serving as a Program Director for the National Science Foundation in the Division of Undergraduate Education. Her education includes a BS in Chemistry from the State University of New York at Albany, an SM in Chemical Engineering Practice from the Massachusetts Institute of Technology, and a PhD in Chemical Engineering from Polytechnic University (now NYU-Poly).

Graduation Celebration—from left to right: Harvey, Vanessa, Pam, William and Heather

My Circus: Please Note That I Have No Formal Training in Juggling

Amber Flynn Charlebois

It was a beautiful Friday afternoon and I had just finished tabulating the responses to the bonus from the Organic Chemistry exam I was grading. I was very much interested in the students' answers to the following question:

Bonus—I am due to have my third child soon; if it is a girl, what should her name be?

(a) Sarah
(b) Victoria
(c) Elizabeth
(d) Pepper

You see, my husband, Jay, somehow got into his mind that the name Pepper was appropriate for a girl. In addressing the stalemate we had reached in coming up with a good girl name, I convinced him to let me poll my organic chemistry class in an effort to gather additional data. In my class of 50 something students, only three chose Pepper. Additionally, many students wrote in things like "Pepper is my dog's name!" which only helped my case. Luckily these results allowed us to take Pepper

---

A.F. Charlebois (✉)
Department of Chemistry and Pharmaceutical Science, Fairleigh Dickinson University, M-SB1-01, 285 Madison Ave, Madison, NJ 07940, USA
e-mail: charleb@fdu.edu

out of the running. It turned out that the winning name with almost 20 votes was Victoria. My daughter's name is in fact Victoria.

It is experiences like this one that prompted me to jump at the chance to contribute to this work. I am personally deep in the trenches of trying to successfully blend two of the most amazing things, motherhood and professor-hood. It is my hope that sharing my experiences will provide not only information, but also insight and hope to the reader. I have discovered that this *blend* is more of a juggling act, rather than a balancing act, as there are very often additional variables that need to be considered. Here are *my* specific roles in this thing/call life; in other words, here is what I have to juggle.

- I am the mother of three beautiful children; Steven is my smart and funny 11-year-old, Matthew is my resourceful and hardworking 8-year-old, and Victoria is my curious and empathetic 5-year-old.
- I am an Associate Professor of Chemistry at Fairleigh Dickinson University in Northern New Jersey where I teach Organic Chemistry, have an active research group, and advise the chemistry club.
- I am a wife to my best friend and fellow chemistry geek, Jay. Yes I use the line, "There is a lot of chemistry between us!" entirely too often.
- I am a sister, friend, mentor, and aunt to some of the most amazing people on this earth.
- I am Dr. Demo: On Site Science, where I am able to share the excitement of chemistry with young children around my community, by doing hands-on interactive science demonstrations at local camps, day care centers, birthday parties, and in the classrooms at my children's schools.

In that order, most of the time.

There are days in my life when I feel like, "I got this!!!!" And that I am doing an amazing job at all of the roles I have in life. There are, however, also some days when I feel like I have failed in all of these roles. Usually, I can confidently say that of my five roles, I have dominated at least three on that given day. Maybe by the time my kids are in college themselves, I can claim I have this Professor-Mother thing mastered, but today I can share with you some of my experiences that can guide or may even inspire you. In this chapter I will tell my story, which I hope is helpful and insightful. My disclaimer is that many of the challenges I share here are the same challenges I continue to work on myself, *still*.

Stop Trying to Control Everything; Sometimes You Have to Just Let Go

My parents were amazing people whom I admire to no end. My father was a self-employed dairy farmer and my mother worked in retail. They both spent most of their lives working very hard so that my siblings and I could have most of what we wanted. So from very early on in my life, I was exposed to a working mother lifestyle and I did not know any different. Growing up, I never contemplated being a stay-at-home mom, not because my Mom worked, but because it was not in my personality. But all along, I knew I wanted to have children. I guess I was a bit naive, but I never spent any time or energy thinking about how I would pull it all off; I just did it.

When I was in graduate school (University at Buffalo, NY), I met Jay, my study partner/drinking buddy turned soul mate. We realized quickly that we had a good thing and so we tied the knot. Both Jay and I were nontraditional graduate students in that we were almost ten years older than most of the "just out of undergrad-grad" students. We had both worked in the chemical industry for several years before we decided to attend grad school. So my biological clock was ticking when we began to think about starting a family, at the same time as trying to find a postdoc appointment. I knew that I ultimately wanted to teach at a primarily undergraduate institution. I thought that would be the best place for me to have both a family and a successful and satisfying career. Therefore, as scientists, *of course*, Jay and I started calculating when to get pregnant so that the baby could be born during the summer between my postdoc (University of Illinois Urbana-Champaign, IL) and starting my first academic position. It was perfect! We were due to have our first baby in July 2002, and I was getting my CV sent out to all these awesome undergraduate institutions and we had timed it all perfectly. Until I miscarried.

WAIT! That was not part of the plan. NO! I had it all set, I was going to get everything I wanted and it was going to be perfect. NOT. I was 36 years old and not pregnant. After getting over the shock of the loss, Jay and I decided to just try again. We never really talked about what it meant and I never really thought about what it was going to be like, but we agreed to just deal with the timing issues of childbirth, whenever it happened. In other words, we were flying by the seat of our pants, but we were doing it together so it felt comfortable.

I targeted my academic job search for areas of the country where Jay, also a Ph. D. chemist, could find a position. Jay was interested in working in industry, so we knew his job search would have to wait until we were closer to the actual relocation date. I interviewed at several institutions and accepted a position in New Jersey, and Jay and I started planning the next chapter of our lives together as we organized the move from Urbana-Champaign. New Jersey seemed like as good a place as any for Jay to find a chemistry job, so we were well on our way.

And guess what, we were successful at the pregnancy thing too and we were due to have our baby right around Thanksgiving in 2002. Okay so wait, that meant that I was having my first baby smack dab in the middle of my first semester ever of teaching. Not an ideal situation, but we could do this, right? As the due date got closer, Jay was still deep in his job search, and I was so very excited about my teaching position. In chatting with HR at my institution, we discovered that there is a requirement to work for 12 weeks in NJ before disability kicked in. My start date was Sept 1st. The middle of November would not be 12 weeks no matter how you counted it. I was not eligible for disability. I had NO sick time accrued to date and Jay still had not found employment. Some members of administration and faculty offered to donate their sick time to me, for the birth of my child, but HR denied this request because pregnancy was a planned event. They did inform me that if the birth had to be cesarean, then I could use donated time. Looking back, I think I was not as stressed as I should have been.

My adorable 8 lb 7 oz bundle of joy, Steven Flynn, arrived three weeks early, on November 3rd. Members of the chemistry department graciously volunteered to cover my classes until I returned. Not one of my classes had to be canceled, and I was back at work after ten days. At that ten-day mark, I was feeling great and as long as I did not cough or sneeze, being in front of the classroom was fine. This was all possible because Jay had still not found his dream job and so he was able to stay home with Steven while I went back into the classroom. As I reflect, I think I got the better deal because Steven was colicky. I was able to kind of escape some of the extra stress by going to work. Have I mentioned yet that my husband is amazing? The whole thing worked out because I only had to be on campus for a little over four weeks and then I had the entire winter break to spend with my new little Steven. Things started taking off for Jay as well with several interviews and finally a great job offer. He started his new job in January, and so with both of us gainfully employed we could both relax and enjoy our new little life.

Funny little side note: HR in doing their job to the best of their ability actually made an inquiry to the Dean when the insurance papers came in requesting me to add my son to the policy. They were concerned because I had not taken any time off of work, and yet I was adding a child to the policy. They even wanted to force me to take time off without pay. I can't make this stuff up.

Almost like déjà vu, we successfully calculated the timing of my next pregnancy, which also did not come to fruition. This time miscarriage was even more difficult for me. I had made it past the three-month mark; I was actually in the 5th month. Ironically, it seemed to me that there was some force not allowing me to give birth during a school break. It took much longer to heal from this one, and to decide to try again. But again we mustered up the energy and the hope to try again and so my second son, Matthew Michael, was born March 1st of 2005 at a whopping 8 lbs 15 oz. It turned out that for this pregnancy, I was able to plan more effectively and was able to develop and teach my first online course. This allowed me to work from home most days once Matthew was born.

No, Really... Just Let Go

My point with these stories is that this idea of "control" is not possible for all things. In theory, planning your pregnancy around semesters of teaching sounds perfect, but may not always happen the way you envisioned. Don't get me wrong; it is worth a try. And I know many examples of summer babies of women faculty, but if the timing does not work perfectly, just make it happen the way it works for you. And you should know that my third child, my beautiful 9 lb 7 oz girl, Victoria Rose, was born June 2nd, so I got my girl *and* I got my summer baby.

But the "control freak" in me comes out in other ways. I had a hard time letting go when I was on maternity leave with Matthew. I thought I was the only one who could teach the material correctly, so I ended up getting back into the classroom pretty quickly after Matthew as well. This is one of the traits I have that I am still working on daily.

I struggle with letting other people help me and with letting go. Two major examples come to mind.

My husband does not fold the towels correctly. Most of the time, Jay does not contribute to the laundry process. It is because he is in charge of all the shopping and all the cooking—*I scored big time, right?*—and I do the laundry and cleaning. That is just how we break it up and it works for us. He does so much around the house, I feel a little weird complaining about it, but it drives me crazy when he folds the laundry. He folds the towels all wrong and they don't fit in the linen closets correctly so I have to refold them. I realize logically that this is just senselessness on my part and a waste of my time when I stop everything and refold. But when I am deep in the daily grind, I can't help myself. I need to remind myself regularly of two things: 1. I cannot possibly do everything myself and 2. When I get the help I need, I have to be open to different interpretations of success. I need to stop trying to control the folding of the towels and let it go. When I catch myself refolding, and think about it, I quickly realize it is often my overall approach to life that needs adjusting.

Second, I have applied for and have been awarded a semester sabbatical for next spring. I am so excited. I have plans to travel and work with my collaborator in Buffalo for two weeks, and in addition, I intend to get lots of writing done because I have recently and seriously realized that manuscripts do not write themselves. I know I'm ready for a productive scholarly semester without the worry and pressure of teaching and service. GREAT! Except... somehow it feels very empty for me. In the back of my mind, I want the department and the university to need and miss me so intensely, that things fall apart when I am gone. I mean, I don't want things to fall apart totally, I just want everyone to say, "It is SO good to have you back," when I return next fall. I want the students to wish they had me teaching them Orgo II and I want the department to feel that the campus recruitment day was just not the same without my enthusiasm and energy. Again stepping outside myself, I realize this attitude is selfish and narrow-minded. I need to come to terms with the idea that everyone is replaceable! And although I put everything into my job, if I am not there, someone will teach my students organic chemistry and some of my students will go on to become chemistry graduate students, and others will even get into medical school. Life will go on, with or without me.

When talking about these feelings with one of my mentors, Kathryn, she shared with me this quote: "The ultimate measure of a leader is what happens in their absence" (LaBranche). Have I been thinking about this all wrong? My leadership and influence on the people in my life and on my students (and even on my children) are to make them better people so that they can thrive in any situation. This quote made me realize that the most important thing I can do is challenge those around me to be the best they can be in any situation and help them find the tools necessary to achieve success. It is not all about me. I am not the be-all and end-all. It is more about the people I influence, encourage, help, trust, and mentor. I am just a piece in their big picture. You know what? I can do this. I can go on sabbatical next spring and concentrate on moving my research agenda forward and not be preoccupied with how the campus is doing without me. I hope I can, anyway.

Trust Your Gut

There are times in your life when something deep inside your being tells you to run away fast. Those internal mechanisms are real and they are important to heed. With that said, I confess that I did not follow my instincts this one time and it completely changed who I am today.

It was at one of the interviews I had in New Jersey. It was a medium-sized state school and I was interviewing to teach Biochemistry in their program. At the very beginning of my individual interview session with one of the more senior members of the faculty, he introduced himself and stated that he was the biochemist and that the new hire would replace him as he was going to retire soon. He looked at my CV in front of him, looked up at me, and said, "Well, it is good that you are a woman; it would be better if you were a black woman, but it is good that you are a woman!"

Seriously, I should have run fast and far, but two-to-three months later, when I got the phone call and the offer, and no other offers had come yet, I accepted it. I had convinced myself that although I did not appreciate this mentality, this person was on the edge of retiring and everyone else in the department seemed fine, and it was a job offer after all. So I accepted this position formally. Ironically, two days later, I got a phone call from a second institution asking me to be patient because there was an offer coming and they were very excited about bringing me on board. My heart sank as I told them I had already accepted a position. So in the end, I honored my acceptance of the original institution and started my academic career that September at the state school that was happy to have me because I was a woman, which brings me to my next cliché.

Should I Stay or Should I Go Now?

After four years at that institution, I tendered my resignation. At the department meeting when my resignation was announced, the comment I heard from the former chair of the department was, "Well that is *great news*!" Let me interpret the meaning of "great news" in this statement. He was happy that I was leaving; he could not care less where I was going or what kind of position I was taking, he was simply happy that I would no longer be at *his* institution. The original interview comment and the final "great news" comment were the two bookends of that chapter in my life. At that time, I was so relieved to have accepted the offer from a fantastic small liberal arts college, also in New Jersey, and was most excited to be getting out of there. The challenges that came with my new position were that I had to take a substantial pay cut and I had to start the tenure process entirely over.

The decision to move was probably the most difficult one I ever had to make. I was at the end of my fourth year at an institution that had great benefits, great students, and *mostly* wonderful colleagues. All indications were that I would be getting tenure that next year. But tenure meant that I would be stuck there forever and deep down I knew that I had to get out, and I had to get out NOW.

When I reflect back on those four years, it is hard for me to comprehend that it was me going through it. I was in the middle of a hostile work environment based on my gender. It was never anything sexual; it was just a constant disregard and disrespect aimed at me and the other untenured woman faculty, Anita. There were two men in the department who were the drivers behind this treatment, and the rest of the faculty in the department seemed to let just it happen. These two men (one was the chair) screamed at us both in meetings behind closed doors and in the hallways while students and colleagues were present. I was even physically threatened in front of students. We were regularly referred to as "the ladies" or "those two", and these two bullies never learned our names and got us mixed up all the time (we look nothing alike, except that we were both women). There is not enough room in this chapter for me to describe the numerous different experiences and situations we were exposed to and there is also no way that I could truly describe on paper the way these painful experiences made us feel. We had each other to both validate and corroborate our experiences and our emotions, but it didn't dull the horror. My morale was depleted completely and self-confidence was nonexistent.

It all came to a head in the spring of 2006, when it was clear that Anita was not being renewed after her 4th year. I still have no idea how this was allowed to happen. You have to understand, she was an amazing teacher, she had publications (including a children's book about the periodic table), she had been awarded an American Chemical Society Petroleum Research Fund (PRF) Grant, and she participated in plenty of service. She was *better* than me. But looking back it is clear she was more of a threat to them; Anita was stronger than I was. So they fired her. It became evident that once her nonrenewal was announced, they were starting to target me. I dreaded department meetings and any interaction with the two men who were so dead set against women being in their department. The other faculty in the department began "standing up for me," at this point, but honestly it was too little too late. This situation had finally started to affect my inner being, my health (my body was starting to respond to all the stress), and my family (my temper was short and I was spending many hours at home crying and sobbing). I was a mess. Even worse, I was starting to believe what they were implying and saying about me and I was starting to really second guess everything in my life. I had to leave, I had to get out. And so I did.

It has been seven years now since I moved to my new position and life is very good. I am an Associate Professor with tenure and I have no regrets whatsoever. My amazing colleagues and I provide each other challenge, respect, accountability, and collaboration so that we can offer our students the best possible learning environment and college experience. This is an atmosphere where I am excited to bring my children when needed. This is a place where the faculty, staff, and administration all support the notion of family and all that goes along with it. This is a place where my children love to visit and where they dream of attending college someday. This is how it was in my dreams.

Bottom line... The motherhood/professor-hood balance is easier to pull off when you are happy and satisfied in your day job.

Choose Your Battles Wisely

The second most difficult decision in my life took almost a year to come to terms with, but finally Anita and I decided to file a lawsuit against the university for A Hostile Work Environment Based on Gender. We felt that we finally needed to stand up for ourselves. Anita lost her job and I lost the salary and the time toward tenure. In making the decision, we especially wanted to be role models to the many young women (and men) who had watched us treated so horribly. If we had not fought back with the lawsuit, then the students would be left thinking that it was okay for women to be treated in that disrespectful and degrading manner. When it was all over and the dust had settled, we threw a Vindication Party to celebrate our success. We invited all the many wonderful people in our lives who helped us while we were experiencing the hostility and for all the people who helped us with the lawsuit process. We made sure we invited all the chemistry majors during our time there to celebrate with us. We had to show them that what they saw us experience was not acceptable, that people were held accountable, and that we had survived.

When we met with our lawyer, Sam, for the first time he asked us what we wanted from the suit and if we wanted our jobs back, to which we replied, "Absolutely NOT!" I told him that I wanted institutional change and the two bullies to be fired. He told me that a lawsuit can really only get you financial restitution, which sometimes may cause institutional change and might cause the university to modify employment terms. At that time, that was good enough for me. By no means did we get rich from winning this lawsuit, but we're content in the knowledge that we caused some institutional change. We have paved the way for any future women to work in the chemistry department and for any female students in that department as well. Most of all I think we were validated. It became public knowledge that we were treated unfairly and without the respect we deserved. This validation was instrumental in my healing process and in my moving forward.

Looking back, one of the things that helped me when I was in the middle of the hostile work environment was that I documented everything. I had a journal and kept track of (with dates) everything and every interaction (what happened and how I felt). Then, when the time was right, I worked with the Dean and followed the proper channels within my institution to file a grievance against these two men.

Since we were unhappy with the outcome of the grievance, we met with Sam and filed the lawsuit, which took more than three years from start to finish and involved many depositions and expert witness' testimonies. The process was long and sometimes difficult, but ultimately it was a cathartic and freeing experience.

Lean on Me

I have to give a shout-out to a program that helped me through this entire decade of my life. I attended an Effective Negotiation Workshop offered through the University of Oregon and ACS Committee on the Advancement of Women Chemists, or COACh. It is a grassroots organization that works to increase the number and career success of women scientists through innovative programs and strategies. More information can be found on their website http://coach.uoregon.edu/coach/.

It was attending this workshop while I was deep in the thrall of the hostile work environment that gave me the strength to keep going. I remember learning how and where to sit in a room to have the most influence and presence. The other thing that sticks in my mind clear as day is when you are preparing for a meeting it is important to be completely prepared, of course, but to also give yourself five minutes just before the meeting to just BE, to mentally prepare yourself. These two very small pieces of insight gave me the strength to go back to my university and sit tall and have a strong presence in every meeting (even though on the inside I was a basket case). It is what gave me the strength to search for another job and move on with my life. Looking back, this workshop was a very instrumental part to my healing and my recovery.

A second shout-out goes to the American Chemical Society's Women Chemists Committee (WCC). Once I became an associate member of this National Committee my life was changed. I was able to network with women and men from around the nation who valued the importance of women in the chemical enterprise. This group also helped me see the important role women had in chemistry and allowed me to help them with their mission, to attract, develop, promote, and advocate for women chemists. I am still so excited to be part of this very important organization as we continue our mission with programming and outreach to the society as a whole. In this group, I found a safe place for me to learn how to be a more effective leader and a better person. I found a new family.

Day to Day: You Win Some, You Lose Some

Being effective in all your roles on a day-to-day basis is not always possible. There are days I have had where I feel like I have failed in all of my roles simultaneously. These days are few and far between, but they do exist. If I wrote this chapter and pretended they did not exist, I would not be being truthful.

I clearly remember this one day when I was not successful in any of my roles. Victoria was almost a month old and she had to have an MRI of her back. The only opening they had (unless we wanted to wait a month) was one Tuesday morning in July. It was the same Tuesday that I had agreed to teach some local high school students a 3-hour laboratory course about biodiesel and other alternative fuels. It was a good source of summer income and was only for 3 h so I had agreed to it back in May. But that meant I could not go to the MRI with my newborn daughter, and instead I would teach this group of future chemists about how to synthesize biodiesel. Of course Jay agreed to take the day off and take Victoria to her appointment in New York City without me. I was not okay with this arrangement; I really wanted to be there with my new baby. I checked and none of the other faculty could switch with me, so I was forced to come to terms with the situation. I started to accept that Jay could do this and so could I.

This dreaded Tuesday finally arrived. To top things off, Steven, my 5-year-old, woke up with a fever and could not go to day care. Jay could not take Steven to the hospital with him as he would be busy with Victoria. So there you have it, I would tote my sick 5-year-old in with me to campus and teach this lab. During the school year when my kids are not feeling 100%, I have been known to bring them with me to campus, and hire past students to babysit during my lectures/labs so that I would not be distracted from my teaching. But during the summer, there were no students around for hire. So I improvised and set my son up on the student lounge couch (just around the corner from the organic lab where I was teaching) with some snacks, some blankets, and the DVD player for the 3-hour lab. I checked on him every 15–20 minutes to make sure he was fine. The secretary and lab technician helped me by checking in on him regularly as well. And, he was fine. He actually did great.

Everything was fine.

But it was not fine with me. I let my beautiful daughter down because I could not be there with her for her MRI. I let my feverish son down because he was not feeling well and I stuck him in a room for 3 hours. I sucked at teaching biodiesel, because I was not totally present with them as I worried about my daughter and even more immediately my son, as I ducked out of the room frequently.

But in reality, everything was fine. Jay did great and Victoria's MRI came back totally normal. Steven's fever went away and went back to day care the next day. And finally, my student evaluations did not reflect my internal turmoil, but rather indicated that most students enjoyed the lab and learned a lot from the experience. Everything was fine.

Everything really was fine.

The Help

However, my feelings of inadequacy on this day were overwhelming. Actually, these feelings occurred whenever any of my children were sick, or there was a snow day, or if school was closed. What the heck was I doing, thinking I could really pull this all off? Something needed to change. I had been toying with an idea my friend and mentor Kelly had shared with me as we discussed work–life balance. She told me that if I could afford to pay someone to help me in my life, that would also make me more effective in the rest of my life, then I should go for it. And I agree. It was time to buy some help. It was shortly after this Tuesday experience that we decided to get a nanny.

There is stuff that goes with the nanny thing. For example, I had to deal with some guilt about having a nanny, and I am sure some people judged me for it, but it was what I needed at that time to maintain my sanity. My first nanny was perfect and our world was a better place with Nina in it. She became part of the family and my kids still talk about her and beg for her to come and visit. My second nanny was not so perfect, but it was still better than no nanny. Ultimately, having a nanny that was just OK made it easier to decide I could do it all on my own again. All said and done, I had a nanny for almost two years and it saved my life. I returned to babysitter/day care mode once Steven was in school, which just so happened to be the fall I was going up for tenure. I had successfully used the extra time in my daily schedule during the 2-year nanny phase to get myself completely ready for my tenure submission. It was perfect timing.

Two Generations of Role Models

My children are with me on campus all the time. Anytime they can come with me they do. They know many of my colleagues by name and they know where the cool trees to climb are located, and they for sure know where the Dining Hall is. They know the code to get in my lab and office and they know where the markers and pens are in my desk. They know where all the outlets are in my office for charging their iPads and DVD players. They know their way around the Science Building and they know which faculty members to avoid (there is one really mean lady in the Biology Department). Most importantly, my children know many of the students. I think it is amazing that my students interact with my children in a positive way and become role models for my kids. Several students have donated their old children's books or decks of Pokémon cards to my children. I just think that interaction is priceless, as my kids may now aspire to be like my students (the good ones anyway, wink wink).

Then it goes the other way too. Most of my students each semester have met/seen my kids at some point. I try very hard to be a positive role model for my students when they see me interact with my children. I love that the young women (and men too) see me balancing my family with my career. I am not saying it is easy or perfect, I am just saying it is doable and it is great fun!!!!

There is a high-energy barrier for me to be professor and mother at the same time. When I have my children with me around my students, I feel torn. I feel like it is hard to be firm, nurturing, supportive, and instructive to both my children and my students at the same time. It almost feels like I have a double personality. My voice changes and it feels like my personality changes too as I go back and forth between the two types of interactions. That being said, without a doubt, I will continue these interactions because I love that my students see me as a mother and, even more so, I love that my children see me as a professor.

From the Mouths of Babes

During my first year at FDU, I initiated a daylong celebration of sustainability on campus that was named Green Day. I did not realize how much it had infiltrated my entire life until the very innocent little voice of my then four-year-old son Steven chimed, "Mommy, I know that tomorrow is Green Day at your work, but when is Yellow Day going to be?"

When I asked Matthew what he likes best about me being a professor, he shared that it was because I teach him about science and that I can come to his classroom and do demonstrations. As Dr. Demo, I go into all my children's classrooms to get kids (especially girls) excited about science. I very much enjoy sharing my love of science with my children and their classmates as often as I can in my busy schedule. I have very rewarding experiences with these kids and I am hopefully acting as a catalyst for our future scientists. When I asked Matthew what he likes least about my job, he told me he did not like when I have to stay at work late at night. Victoria's only comment about my job was that when she comes to my school she is really bored most of the time, unless there is a student hanging out with her and taking her to the cafeteria.

Steven said to me last month while walking on campus, "Mom, it's going to be weird when I come here for college. I mean it's kind of like I grew up here." I enjoy that right now they think they are going to come here for their undergraduate experience, and that would be fabulous as far as tuition is concerned, but somehow I expect my campus is not where they are going to want to be when they graduate from high school. Time will tell.

The Charlebois Family—Jay, Amber, Toria, Matty, and Steven, 2013

Words of Wisdom

I enjoy my job so much that, honestly, most days it does not seem like work. I feel like somehow I am getting away with something and that one day someone will find me out and make me do real work... I feel blessed that I have found a job where I feel happy, safe, challenged, and satisfied. Mix that with my fun and crazy little family and you have my life. Don't get me wrong, there are some days that life is not all calm and peaceful, but the majority of my days are a crazy-busy juggling act that I absolutely live for.

Below are the rules (guidelines) that I had made every effort to live by in effort to achieve my personal balance.

- Find a partner (if that is in your cards) who wants the same things out of life that you do, so that you can save the world together.
- Learn how to say NO! In this insanely busy thing we call life there are things that need to be done, but you do not have to be that person every time.
- Hire people to help you in your day-to-day life, if you have the means.
- Make sure you have mentors/role models in all areas of your life. Identify people who you are like-minded and you aspire to be like. Then utilize their friendship and guidance whenever you need it.
- Stay true to yourself. Always follow your heart. But remember to challenge yourself and step out of your comfort zone once in a while. This is how you will learn and grow.
- Slow down and enjoy your life. (My sister, Patty has to remind me of this one all the time)
- And don't worry about the towels.

Main Steps in Amber's Career

Education and Professional Career

1988	B.S. Photography—Syracuse University, NY
1995	A.A.S. Chemistry—Monroe Community College, NY
2001	Ph.D. Organic Chemistry—University at Buffalo, NY
2002	Research and Teaching Postdoctoral Associate, University of Chicago Urbana-Champaign, IL
2002–2006	Assistant Professor, William Paterson University of New Jersey, NJ
2006–2011	Assistant Professor, Fairleigh Dickinson University, NJ
2011–present	Associate Professor, Fairleigh Dickinson University, NJ

Honors & Awards (selected)

2000   Monbusho/NSF Summer Program at the University of Tokyo Fellowship
2009   Innovative Challenge Award winner for business "Dr. Demo: On Site Science"
2010   Becton College Teacher of the Year Award

Besides her ongoing commitment to the American Chemical Society (ACS) and the ACS Women Chemists Committee (WCC), Amber aims to inspire future science students and chemistry majors with Dr. Demo, an on-site science demonstration for children aged 3–10.

Planned Serendipity

Renée Cole

"I'm a bitch, I'm a lover, I'm a child, I'm a mother …" I could add "I'm a professor" to the chorus of the Meredith Brooks' song. The different aspects of my life each inform the other and make me who I am, and I really wouldn't have it any other way. At different times in my life, the balance among these roles has shifted, but overall I can say that I have been blessed with a fantastic partner, delightful daughter, and an engaging and successful career.

I've often been asked to describe my career path and how I came to be a successful woman in science. During a leadership workshop, it occurred to me that the best answer was "planned serendipity." Many of the opportunities in my life often seemed to me to be the result of good fortune and circumstance, but upon further reflection, are linked to choices I made that put me in the right place with the right connections or skill set to be offered those opportunities. I think this is also true of my journey as a mother and professor.

R. Cole (✉)
Department of Chemistry, W331 Chemistry Building, University of Iowa, Iowa City, IA 52242-1294, USA
e-mail: renee-cole@uiowa.edu

Starting Out

I've enjoyed math and science as long as I can remember. I took every math class I could in high school, and almost every science class. The one class I avoided was advanced biology ... you were required to dissect a fetal pig by yourself, and I didn't think that was something I could stomach. I never have liked gooey, squishy things or anything resembling gore. (Fortunately my husband has always been willing to take care of those tasks that I couldn't handle.) I did have math and science teachers who encouraged me to excel and who supported extracurricular activities in those disciplines. Science fairs and math competitions were something to look forward to, and I usually did well in those competitions. Choosing to participate in these events opened many doors to me, including travel and scholarship opportunities.

My goal from a very young age had been to get a college degree, so choosing to attend college was an easy decision. My goal through most of high school was making sure I had an academic record that would help me get scholarships to fund my college degree. I had the great fortune to attend Hendrix College, where I still have close ties and many fond memories. I started college as a math major, although I didn't really know what I wanted to do other than something in math or science. By the end of my freshman year, I decided I much preferred chemistry to calculus and changed my major to chemistry. Hendrix had (and still has) a great chemistry department where I was encouraged to get involved in research. They also make it a point to take students who are engaged in research to the spring American Chemical Society (ACS) meeting, so I was able to attend meetings in Atlanta and San Francisco as an undergraduate. These experiences helped me see the value of being engaged in the community and had a significant impact on how I mentor undergraduate and graduate students in my role as a faculty member.

Of course, another significant event in my time at Hendrix was meeting my now husband. Our first date was a blind date set up by my friends as part of a social event sponsored by the dorm where I lived. The date went well, and we quickly became a couple. That was in January of my sophomore year of college. Greg transferred to Oklahoma State in August, so we had a long-distance relationship for the rest of my college career. We were engaged in the spring of my junior year and got married the weekend after I graduated from college. It was important to me to have a spouse who was truly a partner, and my "plan" was fulfilled better than I could have hoped.

Unlike many of my peers and the students and postdoctoral fellows I speak to, the decision about the timing of getting married and starting a family was not one that I really struggled with. I grew up in a family and culture where marrying young and starting a family young were pretty much the norm. Many of my peers and colleagues have commented that I got married really young, but it wasn't something that I really thought about at the time. I actually got married at an older age than most of the women in my family, so I didn't think getting married at 22 seemed like getting married particularly young.

My role models growing up were my mother and grandmother, both of whom worked outside the home and changed careers multiple times. Balancing education and finding a career path you enjoy were natural activities to me. My grandmother went back to school to become a nurse when she was 50, and my mother returned to school to complete a master's degree in nutrition while I was finishing up my Ph.D.

Next Steps

I didn't really know what I wanted to do when I started college, but I found activities that were related to teaching were things that I really enjoyed. I had started tutoring students in math and science when I was in middle school, and this continued after I started college. In addition to tutoring, I worked as a laboratory and stockroom assistant. These activities combined my love of science and the joy in helping someone understand a concept or process. By my senior year of college I knew I wanted to be a college professor, and a Ph.D. was a required step to achieve that dream. While at Hendrix, I had the opportunity to participate in two Research Experiences for Undergraduates programs—one at the University of Arizona and one at the University of Oklahoma. My experiences at Arizona and Oklahoma gave me a good idea of what to expect in graduate school, and I never considered doing something else after I graduated. I chose to attend the University of Oklahoma (OU) and continue my work with Roger Frech, working towards a Ph.D. in Physical Chemistry.

Married graduate students were not uncommon at OU, and there were a mix of single and married graduate students in Roger's group. Being married meant that I tried not to work too much at night or on weekends, which meant being more focused while I was in the lab. I still participated in study groups while I was completing coursework, and regularly spent some time in the lab on Saturdays. Most of our friends were other couples from the chemistry department, and we would regularly get together on the weekends to play games or just hang out. Greg often describes himself as a chemistry groupie because he has spent so much time at chemistry-related social events.

In addition to a Ph.D., I also knew that I wanted a family, and my husband and I decided that the end of graduate school while I was writing up my dissertation would be good timing. For most of my friends, it had taken six months to a year for them to get pregnant, but it didn't take nearly that long for us. The result was that my daughter was born about 18 months before I was ready to defend my Ph.D. Particularly at the time, and regularly even now, when people find out I had a baby during graduate school, the reaction is "I can't imagine having a baby while working on my Ph.D." My usual response is that I can't imagine having a baby while trying to get tenure or waiting until afterwards (although I know plenty of women who chose both of those routes). The follow-up message is that there is no perfect time to have a baby in terms of your career. When you and your partner decide you are ready to have a child, you figure out how to make it work. In my case, I was fortunate to have an extremely supportive husband and an understanding graduate advisor.

My research involved solid-state spectroscopy, so there were no particular hazards that I had to avoid to continue doing my research while pregnant. I actually don't know what the university's maternity policy was at the time because I never asked. I was lucky to have a relatively easy pregnancy and delivery, so I was able to resume my work in the lab after a couple of weeks, although with an altered schedule. My husband worked a fairly early shift and was usually home by 2:30 pm. This allowed me to go into the lab in the afternoons for group meetings and to do any experimental work. I could do much of the analysis and writing up of results at home, so the schedule worked for what I needed to do. Once my daughter was old enough for day care, we found a licensed day care home with a wonderful caregiver. I would go into the lab in the late morning and Greg was home early in the afternoon, so this also gave our daughter quality time with each parent. Because I was working as a research associate when my daughter was a baby, it also meant that it was relatively easy to stay home when she was sick or to take off part of a day for well-baby checkups.

Celebrating with Greg and Mackenzie after graduating with my Ph.D. (May 1998)

After I finished my Ph.D., I had to choose between a faculty position at a very small college (I would have been one of two chemistry faculty) or a postdoctoral position. Although I enjoyed my dissertation work in solid-state spectroscopy, I found that I was very engaged by my work in updating the undergraduate physical chemistry laboratory, and I wanted to do more work in this area. I was fortunate to discover chemistry education research (thank you to Palmer Graves and the rest of the Abraham group), but I did not have much experience in this field. This led me to accept the postdoctoral fellowship in chemistry education at the University of Wisconsin, and we moved to Madison, WI, when my daughter was 18 months old. Childcare in Madison was significantly more expensive than it had been in Norman, particularly for a child under the age of two, so Greg stayed home with her until she turned two. Having a husband who was willing to follow me to different positions and to take on the role of running the household definitely made it easier for me to pursue my career goals. Choosing the postdoctoral position was also a very good choice in this regard. I had good mentors who introduced me to the chemistry education research community and encouraged me to get involved with activities on a national level.

Growing Up

When it came time to look for faculty positions, I was most interested in finding a position at a primarily undergraduate institution, preferably in an area of the country that was closer to our families (who were mostly in Arkansas, Oklahoma, and Texas). At this point, my primary focus was on teaching, although I still wanted to stay engaged in some research. I was offered a position at Central Missouri State University, which later became the University of Central Missouri (UCM), and we moved to Warrensburg, MO, in January 2000.

Greg went back to school to finish his degree after I started my faculty position, but he still took care of a lot of the household and childcare responsibilities. Being able to support him in completing a college education was a sweet reward after the years he spent working factory jobs so we could have good insurance while I was going to school. He graduated with a bachelor's degree with a major in secondary education and social studies in 2003, and a few weeks later we bought our first house.

He did a lot of substitute teaching the first year after he graduated, which led to a full-time position teaching seventh grade. It was shortly after he started teaching full-time that I commented one night that life seemed a lot more complicated in trying to get everything done. Greg commented that it was because I no longer had a personal assistant. Shortly afterward we hired someone to come in and clean every other week. That was one of the best gifts we gave ourselves. This meant we didn't have to spend every weekend cleaning the house (and provided an incentive to keep everything picked up).

The chemistry department at UCM was very family friendly, which made the transition to professor and mother much easier. Several of my colleagues had children about the same age as my daughter, and no one seemed to mind if I brought her up to my office. It wasn't long before I had a small loveseat and drawer with books and toys in my office so that she had a place when she came to the office with me. She grew up playing with molecular modeling kits like tinker toys. Liquid nitrogen ice cream was a treat, but not necessarily a novelty for her because it was a regular part of departmental picnics and gatherings. I guess that's one advantage of growing up in a chemistry department. She participated in hands-on science activities when we would do demo shows, workshops, or activities with different schools groups. She was usually with me when I was preparing for Science Olympiad events or classroom demonstrations, so she was engaged in science from a very young age.

I think being a mother also encouraged me to be more engaged with science education activities in the local schools. I was one of the sponsors of the UCM student ACS chapter, and we regularly did chemistry demonstration shows and hands-on science activities. I also became involved with mentoring middle school students for some of the chemistry-related Science Olympiad events. Once my daughter entered middle school, she joined the team, and I became more involved. However, not all the students knew me outside of being Mackenzie's mom. One afternoon, when I showed up a little before the end of practice, a few students were working on an event focused on science terminology. They were struggling with one of the terms, and I corrected their definition and made some suggestions for how to remember it. One of the students looked at me and said, "What do you know? You're not a scientist!" The students who knew who I was and what I did for a living just looked at him with open mouths and then started laughing. I don't think he ever quite lived it down for the rest of the time he was on the team.

I was atypical in my department in terms of my level of engagement at the national level and in how much I traveled. I have attended the spring ACS meeting almost every year since I was a junior undergraduate, only missing a few years during graduate school. As I became involved in some national chemistry education initiatives, I also traveled to conduct workshops or for project meetings. Balancing my travel schedule with parenting created its own challenges. When my daughter was small, she refused to talk to me on the phone and didn't seem to really mind too much that I was gone, although I can't say the same for my husband. As she got older, I think she missed me more when I was gone. This was particularly true as she took higher level math and science classes and wanted help with homework. There have been many times when I've gone back to my hotel room for homework sessions. As technology has advanced, Skype and Facetime make this easier, although it can still be challenging.

For many years, I used a planner at work and we had a family calendar at home. This worked fine until the year I ended up committing to present at a national meeting and then discovered it was the same weekend as my daughter's first dance recital. I missed the recital, but was fortunate to have good friends who could take care of fixing her hair and makeup—my husband wasn't quite up to that task, although he did learn to put her hair up in a ponytail. After that, I kept a single calendar, although conflicts still arose when some work commitments were scheduled further in advance than school activities. These days, electronic calendars that can be shared among all members of the family make it much easier to avoid conflicts, although the timing of when events are announced can still cause problems.

Hanging out at a show choir competition. Long days, but we enjoy supporting Mackenzie's activities (2014)

Moving On

As I moved through the ranks at UCM, my research program in chemistry education grew to the point where I had to make a decision to scale back or move to a new position with more support for research. I loved teaching and my students, but the teaching and service load at UCM were more than I could sustain along with my research agenda. I began applying for positions at research universities when my daughter was in middle school, but nothing worked out for a couple of years. When she was a freshman in high school, I was offered a position in the chemistry department at the University of Iowa. This was a fantastic opportunity for me, but it did give rise to a number of discussions about the impact it would have on my daughter who had spent most of her life in Warrensburg and was not happy to leave her friends. As it turned it out, it was a great move for the entire family.

Reflections

I can't say that I really had role models early in my career for how to balance being a mother and professor. During both my graduate and undergraduate studies, I never had a female professor for math or science courses. I never really thought about it until one day when a visiting scientist commented that I always said "he" when I referred to instructors and mentors I had had to that point. When I started my faculty position, I was the only woman in the department. There were a couple of women in biology, but they had chosen to not have children. I did find good friends and colleagues in other departments, which helped with some feelings of isolation. I also had good friends and mentors in the chemistry community. My interactions with other women faculty with children at American Chemical Society meetings and the Biennial Conference on Chemical Education gave me the opportunity to talk to other women who had similar challenges and rewards. Some of these women have become very good friends through the years.

I think my position as a professor and a mother also made me more understanding of nontraditional students who were trying to balance being a mother and student. Particularly at UCM, I had many students who were single mothers returning to school to earn a degree to make a better life for themselves and their children. I had to bring my daughter to work with me many times when school was out, so I didn't object to students bringing children to class on those occasions as long as they were well behaved and didn't distract the other students. I also counted it as an excused absence when students missed class to stay home with a sick child.

A concern of many faculty, particularly women with children, is finding the right work–life balance. Sometimes I feel like I do a good job of finding balance, and other times I feel like I'm on a merry-go-round hanging on for dear life. Some days it all works well, and others I wonder how it will all work out. Lynn Zettler, who was presenting a workshop for the ACS Women Chemists Committee, recently introduced me to a new model. Instead of the model of finding balance, she suggested it was better to think of tending a garden. The different aspects of your life require different degrees of attention at different times of your life, but it's important to make sure everything important in your life gets enough attention and to watch out to make sure the "weeds" don't choke out things you care about. I think this model is useful in deciding what to commit to and what to say no to. Hopefully it will provide a framework that will help me in the future, particularly when I'm feeling overwhelmed.

There have been many times when I've wondered if I was present enough for my daughter, but a recent event assured me that I've been a good role model if nothing else. Mackenzie was taking a course at the university that required her to leave the high school a few minutes before the end of class. Another student was in a similar situation, but was afraid to discuss it with the teacher. Even though Mackenzie had never had any interactions with the teacher, she went to talk to her on behalf of her friend. When I asked her why she wasn't as intimidated by this teacher as her friends, she replied that she guessed she had grown up watching me and other smart, intense women, so this teacher didn't faze her much.

As I reflect on the past and think about the future, I have no regrets about the path I created. I have an incredibly supportive spouse, a daughter who has grown into a bright, independent young woman, and a career that continues to develop. All in all, the rewards far outweighed the challenges, and I'm looking forward to see where the future leads.

Hanging out with the dinosaurs at the Natural History Museum in New York (2005)

Acknowledgments I have to thank my husband, Gregory Cole, for all his support through this crazy journey. I couldn't have done it without him. I also thank the graduate students at the University of Iowa who read through drafts of the chapter and reassured me this was on the right track, and something they wanted to read more of.

Main Steps in Renee's Career

Education and Professional Career

1992	B.A. Chemistry, Hendrix College, AR
1995	M.S. Physical Chemistry, University of Oklahoma, OK
1998	Ph.D. Physical Chemistry, University of Oklahoma, OK
1998–1999	Postdoctoral Research Associate, University of Wisconsin, Madison, WI
2000–2003	Assistant Professor, Central Missouri State University, MO

2003–2008   Associate Professor, Central Missouri State University, MO
2008–2001   Professor, Central Missouri State University, MO
2011–present Associate Professor, University of Iowa, IA

Honors & Awards (selected)

2009   UCM Fall 2009 Convocation Speaker
2009   Missouri Governor's Award for Excellence in Education
2010   UCM College of Science & Technology Award for Excellence in Teaching

Renee is active in chemical education research, focusing on issues related to how students learn chemistry and how that guides the design of instructional materials and teaching strategies. She is also actively involved in the POGIL (Process-Oriented Guided Inquiry Learning) project at a national and international level.

Mother and Community College Professor

Elizabeth Dorland

When do you decide you want to be a chemist or to teach chemistry or to be a professor? I wasn't sure what I wanted to do even after arriving in graduate school. Then suddenly I did. But that's the middle of the story, so let's go back to the beginning.

As a kid I imagined being married and having a family in the 1950s Midwest norm of around two boys and two girls. Very few women in our small town worked outside the home. Most married soon after high school and had families immediately. By the time I got out of college and into grad school in the early 1970s, I had settled on 2 kids as a good number. In that later era of women's liberation, I also thought that it would be a good idea to be at least 25 when I married. This was in stark contrast to the conditions in my small town.

Both of my grandfathers were Nebraska farmers. In the 1950s I lived on a farm until I was 8 years old. The closest I came to doing chemistry back then was making mud cookies to bake on warm rocks in the sun. Or maybe the time my sister and I tried to walk (then run) through a cloud of anhydrous ammonia leaking from a fertilizer tank to get to our house from where the school bus stopped at the bottom of our lane. That didn't go well, so we took a long-cut around through the cornfields.

I lived in the same town as both sets of grandparents. Neither my mom nor my grandmothers worked outside the home when I was young. However, both of my grandmothers had been teachers before they were married (in their day, married

E. Dorland (✉)
Washington University, Campus Box #1138, St. Louis, MO 63130, USA
e-mail: dorland@wustl.edu

ladies had to quit), and both put a high value on education. My parents both attended college, but were married at age 20 and didn't finish at that time. Still, I always knew I would go to college. It was just assumed. My mother went back to college in her 30s and became a teacher too. It seems to be in the blood.

There were no chemistry kits in my childhood, and I set off no explosions. In my small town of 1,500, there was just one room of students for each class from the first to the eighth grades. All of my teachers were female until I reached freshman General Science. I don't recall much about my science education in the early grades, but I definitely remember seventh grade. Miss Bowers filled a large beaker with table sugar. Then she poured concentrated sulfuric acid into the beaker. Readers who are chemists know what happened next. A tall, fat black column of porous carbon grew out of the beaker and towered over her head. We were amazed. That may be the only thing I remember from the whole year.

I had an excellent (male) science teacher when I was a freshman in high school. He had been a farmer for many years, but had gone back to college to become a teacher. I thought that chemistry in particular was interesting. Although we learned about atoms in general science, what the juniors were doing in their chemistry class looked a lot more interesting. Strange and horrible smells came out of the room. They heated mercury and got some red powder. I couldn't wait until my turn came in a couple of years.

But fate intervened, and my mom, a single mother with three kids who had gone back to college to become a teacher, married a fellow student who also had three kids. Our blended family moved to an even smaller town in Kansas with a population of 250. The high school had physics, but no chemistry class. None.

The physics teacher (also the Principal) was excellent, and I did well in spite of his doubts that I would survive as the only female in the class. I always say that I chose chemistry because I had missed out on it in high school. When it came time to choose a college major, I read the descriptions in the catalog. Chemical Engineering! Plastics! Why not?

At this point I had no role models whatsoever for a college-level academic career, and no idea at all about what I wanted to be when I grew up.

At Kansas State I did well in general chemistry lecture. I didn't really like my lab classes, but being the only female major in chemical engineering major wasn't much fun either. So when the chemists offered me a summer job if I changed majors, I bought in. During my college years I worked for a summer at Kansas State and then two summers in the research labs at Kodak Park in Rochester, NY.

Travel was not something my family did. I had seen the ocean only once, on a Christmas trip with a classmate during college. I barely knew what a bagel was. Visiting New York City on that trip and later when I worked for two summers at Kodak convinced me that I would love to live in the big city. So I applied to Columbia, University of Chicago, and as an afterthought, to UC Berkeley.

In fact, going to graduate school was partly a way to avoid taking a job as a technician in a chemistry laboratory. I had no college role models in my family other than teachers. I didn't even know what graduate school was until I met my grad student TAs in freshman chemistry. In my hometown as a kid I remember gossip about the son of a prominent family who was so irresponsible that he was still a student in his late 20s. I later realized that he was a graduate student! That was not a recognizable beast in small-town 1960s. Being a TA would pay my way, so off I went, but not to Columbia.

In the end the attraction of California, where I had never been, was irresistible. I was 21 years old. That turned out to be a good choice for many reasons. My relatives were worried about me driving alone, and by coincidence, the son of a lifelong friend of my great aunt in Nebraska was also going to Berkeley in Chemistry. I asked him to ride along. Guess what? He did, and we got married in Golden Gate Park a year later in front of our grad school friends. We celebrated our 42nd anniversary this year. But that is getting ahead of the story again.

During my first year at Berkeley I was assigned to be a TA for freshman chemistry. As I recall, the course structure included a 1-h recitation session and two 3-h labs per week. My first obstacle was my shyness and my fear of speaking in front of any group. My parents and grandmothers were high school teachers, but I had no intention of following in their tracks. I never intended to become a teacher of any sort.

However, explaining the chemistry concepts in recitation seemed to come naturally to me. The faculty member assigned to my section complimented me on my teaching. Students liked me. Still, giving lectures in front of an audience (the only way science was taught back then) was not something I could picture myself doing. In fact, I was so terrified to give my first presentation in front of my new research group that I cried during a practice run with my future spouse. Amazingly, my advisor and lab mates complimented me on the clarity of my talk. And it was fun.

In grad school my students, peers, and professors found my presentations and explanations to be very clear. I enjoyed explaining scientific concepts very much. Lab work did not interest me as strongly. I continued as a TA in general chemistry at Berkeley, and later in introductory organic chemistry. During the following months I continued to find lab work rather tedious. I started hearing about the community college system in California and the possibility of teaching college chemistry with a master's degree. By the end of my first year, I had decided. I got my M.S. in organic chemistry the next year and started looking for a part-time job to test the waters while my by-then-spouse completed his PhD at Berkeley.

In 1972, I got my first position. It was half-time with full benefits, teaching two sections of introductory chemistry with labs at Diablo Valley College. I was their first-ever daytime adjunct. That quickly changed, and colleges all over the country started hiring part-timers to save money. No more benefits. Luckily, my spouse had access to benefits via his postdoc position and later from his professorships. At that point I did not know how many years it ultimately would be until I had my own tenured position.

I have never regretted my decision to teach in community college. If I had known how long the pathway to a tenured, full-time, and permanent job would be, I might have hesitated. I'm glad I didn't. After finishing grad school I taught part-time at two colleges in the San Francisco Bay Area and as an associate instructor back at Berkeley. Later on I taught at two different community colleges in the Seattle area and at the University of Washington. Later still, we lived in Amherst, MA, and I taught at a nearby small college in Springfield. Ultimately, we ended up in Arizona. But I'm getting ahead of the story again. Perhaps you are wondering when the mother part will begin. We will get there eventually. As you will soon see, travel (both with and without your children) is one of the best perks of being a college faculty member.

When my spouse finished his PhD, we decided to teach overseas for 2 or 3 years. At that time he was thinking of looking for a small college teaching job when we returned, and I wanted a full-time community college position. We decided on the American University of Beirut in 1975 and had a great 5-week trip hopping from Japan to Hong Kong to Thailand, India, Nepal, and Afghanistan. Unfortunately, by the time we reached Beirut in September, the war had begun. We stayed there for 6 weeks in a great apartment with a sea view, but were never able to begin teaching. Hearing gunfire every night and watching documents being burned on the roof of the nearby British Embassy were the last straws. Faculty members and their families were evacuated eventually to Greece to wait until after Christmas to begin the school year. We spent an enjoyable 2 months in Athens and traveling the countryside, but by Christmas the fighting had not abated.

We returned to the USA. I took a sabbatical replacement position with my former employer, Diablo Valley College, and my spouse became a temporary postdoctoral fellow for his Berkeley PhD advisor. At the end of the spring, he was offered a postdoc in biochemistry at the University of Washington in Seattle. I auditioned for a part-time position at Seattle Central Community College and landed the job. During the 3 years we were there I also taught at Shoreline Community College and in chemistry at the University of Washington.

When we left Seattle I was in my late 20s, and we discussed when to start a family. I was very worried about never getting a permanent job and wanted to wait until I had a permanent position before having kids. However, the most favorable job offer my spouse got was nowhere near any community colleges. His least favored offer was. The middle course for him was near at least some small colleges, so we both compromised. After I got a permanent position, then we would discuss children. I should mention that beginning our relationship in graduate school meant that we both had the same demanding schedules. Neither of us was much of a cook yet, so we shared all of our household tasks, including cooking and cleaning. We always have. We each have our strengths and preferences, but we always negotiate. Work out a fair system early on, and be flexible.

I did get a position teaching laboratories and supervising the stockroom at American International College right away, but it was not what I wanted to do permanently. But before I knew it I was well over 30, and another compromise seemed wise. Our daughter was born when we had been married for 12 years. During her first 3 years I taught half-time at AIC. My spouse spent his first sabbatical at his home campus the semester after our daughter was born while taking care of her in his office 2–3 days a week. The second semester she went to day care on the days that I worked. I was not on the tenure track during her early years, and that made time management much simpler. But I was very concerned that I would never be able to achieve my goal of a full-time tenured position at a community college. There are always trade-offs.

In the end, I worked half-time from when my daughter was born until my son was 3 years old and she was 7. I worried all those years that I would never find a full-time tenure-track job, but working part-time turned out to be a luxury I could afford. If I were doing it all over and a position came up sooner, I know I would take it just because there are no guarantees. But I was very lucky. Moving to Phoenix was also lucky, because there were far more community college opportunities in the western USA than in western Massachusetts.

When our daughter was 3 years old, my spouse accepted a position at Arizona State University. He had discovered that he really liked having graduate students and a full research program. I was also somewhat optimistic about my prospects because the Phoenix area is home to the largest community college in the country. I immediately started teaching general chemistry lecture and lab at Glendale Community College about 45 min drive from our home. I had child duty during the days, and I left for the GCC campus as soon as her daddy got home. A year later our son was born in late August, and that fall semester was the only time I didn't teach during the 14 years since I finished grad school. I went back to teaching evenings, eventually at Mesa Community College just a few blocks from our home. When my son was 3 years old I was offered a one-year-only sabbatical replacement position at MCC. My son was old enough for their Children's Center and my daughter started kindergarten. The following year I got a tenure-track position at Glendale, and my son was able to continue at the MCC Children's Center. I dropped him off on my way to GCC. I transferred back to Mesa after 3 years in Glendale to be closer to home. Two years later, I received tenure.

In my view, sharing childcare and household duties is something that is simply nonnegotiable for two professionals. We still share the cooking duties. After our daughter was born, we hired a college student to clean house once a week. When we eventually could afford it, we hired regular cleaners. We never stopped. Do it!

I was very worried that having kids too soon would interfere with my ability to have a fulfilling career. In those days, there were few role models who had successfully combined both. In the end, I taught as an adjunct in community colleges and universities for 10 years before deciding that it was time to have kids, or it might be too late. After getting my MS in 1971, I taught for 4 years in California, 3 years in Seattle, and 3 years in Massachusetts. My daughter was born in 1982, and my son in 1986. Between the two, we moved in 1985 from the east coast to Arizona. I taught part-time and as a full-time OYO (one year only) in the Maricopa Community College District in the Phoenix area for an additional 5 years before being hired full-time at Glendale Community College in 1990.

If someone had told me in 1972 that I would not have a permanent full-time position until 1990, I'm sure I would have freaked out. But in the end, it worked out just fine.

The fact is full-time community college teaching jobs were just as hard to come by in the 1970s and 1980s as university positions. Things finally opened up, at least in Phoenix, when the initial hires from the 1960s began to retire. In the end, I'm not sure that I would have found a full-time permanent position any sooner whether or not we had kids. Being a 2-body problem family perhaps had an impact, but even then there were no guarantees. But I love teaching, and could never see myself doing anything else once I figured that out early on in grad school.

One of the most valuable perks of being a professor is the opportunity to become a part of an international community of like-minded scholars. Getting help from and forming relationships with other faculty are important for both job satisfaction and career advancement. Teaching daytime classes during most of the years I was an adjunct allowed me to do this, but my disciplinary connections (in those pre-internet days) were mainly local. Once I had a full-time job, I wanted what my spouse had since graduate school—trips to professional conferences and a network of colleagues around the world. Luckily the community college district that hired me provided monetary support and encouraged these activities.

It wasn't long before my chemical education network began to expand and to inform my teaching. One of my first conferences was the 1992 Gordon Research Conference on Science and Education, the first-ever education GRC. Chemists at that meeting subsequently organized the first Chemical Education GRC in 1994. It is still going and now is known as Chemical Education Research and Practice. On the bus from the airport to the BCCE (Biennial Conference on Chemical Education) at UC Davis in 1992 I shared a seat with Kurt Sears, who was then a chemistry program officer (rotator) at the National Science Foundation in the Division of Undergraduate Education. By the time we reached the campus, he had already invited me to be on an NSF review panel. Later that week I met Susan Hixson, DUE Chemistry Program Director. Susan also invited me to review, saying that they needed reviewers with a community college perspective. My network was beginning to form. These conferences and panels and the friends and colleagues I met there laid the groundwork for some of the most rewarding activities of my entire career.

The Macintosh SE-30 computer in my office was already connected to e-mail in 1990 and to the campus network, but not much more. But in early 1993 I joined the recently established chem-l listserv and my love affair with online connections began in earnest. That day my Chem 101 students were extracting red cabbage juice to form an indicator solution that exhibits a rainbow of colors depending on pH. As I wandered back to my office to fetch something, I was wondering what kind of molecules were involved. I decided to ask online and sent a quick e-mail to the list. When I checked my e-mail at the end of the lab, I was amazed to find not just one but three detailed responses to my query. Remember, these were the days before Google and web browsers (let alone Facebook), so this was about as amazing as it could be. Chemist friends from chemed-l became in-person friends when we had our first face-to-face meet-up at the next BCCE at Bucknell University in 1994.

From my new contacts I learned about the availability of visual chemistry tutorials and software programs that illustrated reaction mechanisms step by step from CD-ROMs. I became a pesky early adopter who was always pressing my school and department to buy these cool toys for us to use in lecture, in the lab, and in the campus computer lab. With the advent of the World Wide Web, the chime plug-in allowed students to view and rotate molecular models on-screen by using files from the Protein Data Base. I could create my own (horribly ugly) web pages and share my learning adventures with other chemists by passing along the links on chemed-l. Perfect synergy.

Travel is one of my very favorite things, and being a professor provides many chances for adventure. Do not pass them up! In 1999 we were able to go on sabbatical for 6 months in Australia. One of my new chemistry contacts helped me to arrange a position at the University of New South Wales, while my spouse did some research with a colleague at the University of Sydney. Our kids were in what the Aussies call Year 7 and Year 11, and had a great experience attending a local high school. They learned that our Math is their Maths, and that our Sports are their Sport. Surfing was one of their choices for sport, and I think they were crazy to pass that one up. But they did have a fantastic time. We negotiated with their school principals ahead of time and ultimately their credits transferred back with only a few glitches.

In the spring of 2003 I got a phone call I'll never forget. A community college-based chemistry program officer at NSF invited me to interview for the rotator position he was about to vacate in Washington DC. At my interview talk I spoke about my various classroom and lab teaching experiments, including the student video project that was going on in my qualitative analysis lab. I was hired. My son was a junior in high school, so I was reluctant to be gone from home. However, he was very supportive and I promised to stay only one year and to be back for his senior year. My husband was department chair in his university chemistry department at that time and could not take leave, but he also agreed that I should not miss this opportunity. In the end, my college age daughter moved with me and took some classes. I went home or my family visited us about once a month. Many of the other DUE rotators and program officers are still friends.

I gave many talks on NSF programs and returned home with a much wider perspective on educational issues. I was sorry that I couldn't stay a second year and use the vast amount I had learned, but it was a good compromise. I was appointed to several NSF Project Advisory Boards and I continue to review frequently for NSF. Take advantage of this kind of opportunity and of all of the sabbaticals that are available once you are full-time. They create fantastic family stories and are opportunities that should not be missed.

The longest trip I took during my NSF experience was to the 2004 ICCE (International Conference on Chemical Education) in Istanbul. My talk about NSF programs inspired an audience member to tell me that his country needed an NSF. I innocently asked where he was from and he said—Iraq. Amazing. I didn't know it at the time, but contacts I made at that meeting would lead to some great adventures over the next 10 years. One of the keynote speakers was to be the chair of the Gordon Research Conference on Visualization in Science and Education in 2005. I had always wanted to go, but a flight to England was beyond my travel budget. When I mentioned this, he said: "I think I can help." And he did, with both travel and registration support. When I returned to the GRC conference at Oxford

(England) in 2007, I was elected as vice chair for 2009 and chair for 2011. As far as I know, I am the only community college faculty member ever to be a GRC Conference chair.

Liz and her family visiting Africa in 2007. For left to right: Sam, Bob, Liz and Larissa

My science and education conference and NSF activities ultimately created just what I wanted—a fantastic international network of colleagues, friends, and collaborators who work in science education and visualization, and more recently in educational technology, online communication, and virtual worlds. I have traveled to over 30 countries since 1972. Our kids both took their first trips to Europe at age one and remain inveterate and enthusiastic travelers.

As far as career influencing my family, I think that the kids were both proud of having "academics" as parents. The extensive travel opportunities tied to academic conferences and the extra time my summers off provided were both very good for all of us. To this day, we love traveling as a family, and try to schedule at least one trip per year to somewhere new and exotic. We plan to continue that tradition even after the kids have their own families.

I asked the kids recently about the best part of having professors for parents. My daughter says that she particularly loves the travel opportunities provided by family vacations before or after academic conferences in exotic places. Some of her favorites include Australia, New Zealand, Fiji, Africa, the UK, China, Prague, and Rome. Says my son, the family wit: "The best part was being able to build dinosaurs out of their molecular models."

Liz and her children, Larissa and Sam, on a trip to China 2010

I think that making family choices is a very individual decision. Also, the times (and the rules) have changed for the better in many places. Partly my thinking is influenced by my experiences, and partly by observing grad students in my husband's and other professor's groups over the years. When I was a Berkeley grad student, none of the female students had kids either before or during their graduate years. I didn't see how they could. Most were unmarried. The "conventional wisdom" among male faculty (and probably some students as well) was that females were just going to drop out to have families either during or after obtaining their PhD. Only five of us were female in a class of seventy beginning PhD students. More than one of the professors questioned why I was there.

My advice is to choose an institution in a setting that provides the necessary support and resources, and to choose your spouse carefully. And I'm only partly kidding. We have always shared household duties and chores, including cooking and childcare. Negotiate a schedule, continuously revise, and let it evolve. One thing our kids missed was having grandparents close by. If your or your spouse's parents are true gems and live in a great place, consider the benefits if you can.

I believe whether kids and grad school (or a new professorship) can be balanced comes down to how organized and efficient the individual is. But then again, finding success as a professor at a major research university requires those same skills whether you are male or female. Balancing a family with a community college teaching career certainly is easier, but that's not why I chose it. It's more like "it chose me," and I loved it.

Main Steps in Liz's Career

Liz's teaching career began when she was an undergrad TA in Kansas in the late 1960s and then expanded over more than three decades and six states. After graduate school at UC Berkeley, she taught chemistry in two liberal arts colleges, three state universities, and seven community colleges, including 21 years in the Maricopa Community College District in Phoenix and her time at the National Science Foundation in Washington DC. She preferred Community colleges, as these diverse environments provided her with a broad view on the nature of chemical education and its evolution over the years. Liz's engagement with the research community around visualization in science and education has given her important insights into the nature and process of learning. Since 2006, her flexible schedule at Washington University has permitted Liz to continue reviewing for NSF and to attend and present at conferences with multiple perspectives on educational reform, collaborative online communities, and the effective use of technology. Liz hopes to continue teaching and learning for a long, long time.

Chemistry in the Family

Cheryl B. Frech

As a girl with no brothers growing up in the 1960s, I got to participate in both traditional boy and girl chores and activities: mow the lawn *and* weed the garden, clean *and* cook the fish from a fishing expedition, and play with Barbies® *and* with Lego®. My father worked as a synthetic chemist and frequently came home smelling like organic chemicals, or more frighteningly now, the diborane gas that was piped into one of his labs for research on rocket fuel formulations. He sometimes took me to work with him on weekends. I delighted in one display of brightly colored inorganic compounds and another of the more pastel hues of the displayed rare earth compounds. By the time I was in high school, I was often one of just a few girls in my chemistry, physics, and advanced math classes. Yet not one person told me that the study of science was not for girls. So when I enrolled in college, I selected biochemistry as my major at Oklahoma State University.

A fairly typical undergraduate, I took my classes, studied, and had fun. My 17-hour first semester included one hour of marching band, which led in the second semester to a stint in the wrestling and basketball pep band and participation in

---

C.B. Frech (✉)
University of Central Oklahoma, 100 N. University Drive, Edmond, OK 73034, USA
e-mail: cfrech@uco.edu

concert band. My sophomore year I joined a sorority and continued with concert band. While no one in my science classes discouraged me from pursuing my major, I was more actively discouraged in the sorority. When it came time for officers to be selected, I was slated only for positions that required a lot of drudgework: service projects chairman, reference chairman. When I inquired about seeking more of a leadership position, I was told that would not be possible since I was "never around." Around where, I thought? The television room, where girls congregated in the afternoon to watch soap operas and smoke? That was when I was in the laboratory for quantitative analysis and organic chemistry. I had missed a couple of days during initiation week, but that was because the concert band toured the state visiting high schools. Despite this less-than-optimal situation, I stayed in the sorority and graduated in four years with a respectable GPA and headed to graduate school in chemistry at the University of Oklahoma (OU).

When I arrived at OU in 1981, there were no women on the faculty, but just the year before, the Department of Chemistry had begun an initiative to recruit outstanding women for the graduate program. Waiting for me were a small group of women who were one year into the Ph.D. program and who would become my peer mentors and friends. While our undergraduate and high school friends had gotten married and started families, or gone off to work in banks and offices in their women's suits, shoulder pads, and silk scarves arranged like neckties, we were wearing lab coats and immersed in stopped-flow apparatus, rotovaps, and nuclear magnetic resonance. We were required to spend at least one year as a teaching assistant. I found myself at 21 years old (and looking younger) standing in front of my first general chemistry laboratory section and looking completely indistinguishable from the students. In order to maintain order and to be recognized as the instructor, I realized I would have to dress the part and be a little bit strict, at least at the beginning of the semester. That was the origin of two habits that I have maintained throughout my career: always dress like you are the instructor (because you are) and establish some rules for your classroom (you can always ease up later in the term).

Throughout graduate school, I excelled at teaching and at presenting the seminars that constitute part of your degree requirements. Members of my graduate committee would say, "You explained that concept very well. Have you considered a teaching career?" I would always answer no, since I fully expected to work for a chemical or petroleum company, which is where my father worked and where my graduate school colleagues were finding jobs upon graduation. I met my husband in graduate school. Roger was already on the OU chemistry faculty, had been married previously, and had children who were in their teens at the time. We were married shortly before I completed my doctorate. I was able to complete one postdoctoral fellowship at OU before we moved to Mainz, Germany, where we both worked at a Max Planck Institute: he on sabbatical as an advanced researcher and I completed a second postdoc.

By the time we were in Germany, my biological clock had triggered, which in turn initiated many conversations about the possibility of us having children. Roger had entered the university system as part of the post-Sputnik science boom generation. Almost exclusively men, they toiled away in the lab, wrote papers and grants, taught their classes, and traveled to conferences and meetings. Most of them had wives at home who managed the day-to-day tasks of the household and essentially raised the children. We knew if we had children, the dynamic had to be different, mostly because I could not be that kind of wife.

When we returned from Germany, I was hired for two years as a visiting assistant professor at OU, teaching very large lecture sections of general chemistry. I didn't know there were better teaching situations than 250 students in a lecture room, but I loved teaching, and started to work to bring more active learning into the classroom. In 1991, a permanent tenure-track position was advertised at a regional university in the Oklahoma City metropolitan area and I was encouraged to apply. Central State University, soon to become the University of Central Oklahoma (UCO), was hiring someone to teach general chemistry and serve as the general chemistry coordinator: it was if the job description had been written for me. The only downside was the 37-mile one-way commute from our home in Norman. I sought advice and someone asked, "If there was no commute, would you take the job?" I said yes, and he replied, "then take it anyway." I have now been at UCO for 23 years, commuting daily from the southern edge of the Oklahoma City metropolitan area to the northern edge. When I first started commuting I listened to audiobooks on cassette tape to cope with the 45- to 60-minute drive. The technology has changed from tapes to CDs to downloads on my smartphone, but I am hopelessly addicted to listening to books and am never without one in the car. People inquire about the stress of the commute. The time and distance intervals of the commute have been an excellent buffer between university and home concerns.

A few women taught in the UCO Chemistry Department before me, but none were in tenure-track positions. The year that I was hired another woman who already had a young child was also hired. Some faculty in the department had older children and some had grandchildren. In my second year at UCO I became pregnant. This was a pregnancy with major complications, and I missed several weeks of the fall semester after emergency surgery. My department chair treated this absence as he would have any other illness. As my due date neared, I planned to take off the entire spring semester and return to teaching the following fall. The state of Oklahoma offers no parental leave, so I was required to take an absence without pay. (This will no doubt be factored into my retirement calculations at some point, so there will be a financial repercussion.) Our daughter, Alison, was born in February 1993.

When I returned to work at UCO in the fall of 1993, we hired an in-home caregiver. Wanda was an older woman with a lot of experience in childcare who delighted in caring for Alison. I was still nursing Alison that semester, and was able to nurse her in the morning before I left, in the evening when I returned, and before she went to sleep. Neither of my children would ever take a bottle, so Wanda taught them to use a sippy cup from an early age. By the spring semester, I had weaned Alison so that I could resume travel since I had recently been elected chair of the local section of the American Chemical Society (ACS) and needed to attend a training conference.

My only sister is eight years younger than me. We were never close until we both became adults. I vowed that if I had more than one child, they would be close in age so they would have the potential to be closer friends and siblings. And so I became pregnant again when Alison was just a year old, and our second daughter, Emily, was born in April 1995. After her birth, we needed a little longer time as a family, so I took off another semester and returned to teaching in the spring 1996 semester. We again hired Wanda to care for two small children 4 days a week. On the single day that I was not teaching, I dropped the girls off at my parents' house in Oklahoma City on my way to Edmond, so that the girls could get to know their grandparents.

Because my work and home life are geographically separated, my children were not as present in the UCO Chemistry Department as other children might have been. And as such, I was not perceived as a parent perhaps as much as others who had to bring a sick child to work with them, or who had to leave suddenly to pick up a child from school or day care. Of course, this would not have been possible without Roger in Norman to respond to emergencies and to fill in when Wanda was unavailable. I am grateful that he was at a point in his career and his life where he could be the responsive and available parent when I was often an hour away.

The University of Central Oklahoma is a primarily undergraduate institution (PUI) and we do not have a graduate program in chemistry. When I was hired in 1991, the university was not far removed from its normal school/teacher's college roots. UCO faculty have always had a heavy teaching load, usually four courses per semester. Although the department has good instrumentation in chemistry, we have limited laboratory space for research. It's only been in the past year that we have carved out shared space for faculty to conduct research outside of teaching laboratories. In order to progress through the promotion process, I have participated in scholarly activity such as writing and reviewing chemistry teaching materials, writing invited book reviews, and serving as an associate editor for a chemistry journal. Being able to focus on these sorts of scholarly activities has been far less stressful for me than for my peers at research universities.

As the girls grew and started school, Wanda moved on to other babies and we hired college students to pick up our girls after school and stay until Roger or I got home. I never felt terribly guilty about not being around my children constantly. I completely subscribe to the "it takes a village" approach of raising children. Each of the women who cared for our children brought something to our daughters' lives that we could not. Wanda had an excellent singing voice, and had sung extensively to the girls. Two of the college students, Auva and Tonya, were excellent at crafts, and taught the girls how to make simple decorations for their rooms. Michaela was an engineering major who studied calculus on the couch in the afternoons: Emily was in awe. Today there is tangible evidence of the multiplication of these gifts: both Alison and Emily have worked in childcare and served as caregivers for other families.

At UCO I have always made a point to tell students in my classes that I am a mother, and to let them know that when I am not at UCO, I am doing the things that their mothers might have done: going to dance recitals, attending softball and soccer games, and serving as a band parent. Our university has a large population of commuter students, many of whom attend school while raising their own children. Knowing that their instructor is also a parent makes it easier for a student-parent to let me know that they have missed class because of a doctor's appointment for their sick child: otherwise I might never find out the reason for the absence.

When I arrived home in the evening the four of us would prepare a meal and eat together. Over the years the girls heard many dinner-table conversations about our students, departments, and chemistry, and perhaps it prepared them well to face their own education head on. Even though we are both educators, unlike some of our daughters' friends' parents, we did not intervene in our daughters' public school education. If either girl had a teacher they did not particularly like, we told them to stick it out, rather than go to the school and demand a change. The girls learned early on to take responsibility for their own education issues. Emily was only in sixth grade when she realized that she was playing on a competitive soccer team while also enrolled in a rather lame physical education class at school. Figuring she had enough exercise playing soccer, she went to the office and arranged a schedule change and became an office aide for the rest of the year. Surprisingly, the school never contacted us for our approval.

When the girls were 5 and 7 both Roger and I applied for and were granted sabbaticals, and our family moved to Scotland for a year. Roger worked on various research projects at the University of St. Andrews and I taught two chemistry tutorials and worked on several writing projects. We lived in a small fishing village and I was able to work mostly from home and be there for the girls before and after school in a way that I could not in Oklahoma. This was a delightful respite and a cherished year for all of us.

Shortly after we returned from Scotland, I was elected Chair of the UCO Chemistry Department. The previous chair was male and single, and lived only a few miles from the university. Again, I told my colleagues that I could perform this job, but I would do it differently than my predecessor. In fact, the previous two department chairs had spent many nights and weekends handling work in their offices, but I knew I would not be able to do this, and be present for my daughters, who were 9 and 11 at the time. My approach was to organize the work and take what needed to be done home to complete after the girls had gone to bed. I also strove to delegate tasks to get more people involved and to mentor new faculty in projects that benefitted the department. I ended up completing two four-year terms as department chair. During this time our department grew in number of faculty members, credit-hour production, and chemistry majors, we completed several assessment reviews, and we modified our curriculum: a successful run, at least by my account.

In the department I heard minimal grumbling about my commitment to UCO in my role as a multitasking mother, professor, and chair. However, I served under one dean who did question my availability. When my daughters were in elementary and middle school, my schedule was such that I could drive them to school and then drive the 37 miles to the university, arriving between 8:45 and 9:00 before my classes began at 10:00 am on Monday, Wednesday, and Friday. I was never late to class. I was not, however, around at 7:30 am when this dean sometimes wanted to speak to me. During an evaluation conversation, he mentioned that I was not always available when he called my office. I responded that I was always available by mobile phone, and that I was on campus on a regular schedule. He continued to push, blustering about how I might be needed to respond to a vague "incident" that might occur someday. I kept asking, "What, in particular, have I missed up until this point?" He had no answer, and I kept taking my daughters to school until they were able to drive themselves.

In addition to my role at UCO, I have been professionally active in the American Chemical Society (ACS), at the local, regional, and national level. I have met and worked with a number of outstanding chemists through the ACS. One of them is Helen Free, now in her 90s, who, among her many accomplishments, was President of the ACS and has been recognized for inventing diabetes test strips. I served on an ACS committee with Helen when she was well into her 80s and she still attends national meetings. Her husband had children from a previous marriage, and they had six additional children together. I once asked her how she had seemingly "done it all." She replied that it was perfectly acceptable to pay for someone to do the things that you do not have the time to do or that you do not enjoy, such as cleaning house or cooking. What a relief! I immediately hired someone for biweekly housecleaning.

Being a professor and a mother is a challenge that will be unique for each woman's family, career, and institution(s). There is no one "right" way to be successful. Strive to let go of being perfect, or at least perfect in all aspects of your life at the same time. You will miss a few of your children's events, and you probably will not be as fit, relaxed, or current with your hobbies as you might have otherwise been. Don't forget your spouse/partner: they are an integral part of the equation and they will still be around when the children make their way into the world.

Cheryl Frech, *left*, with Roger, Emily, and Alison Frech.

Interview with the Author

1. **How has deciding to start a family or having a family influenced your career? How has your career influenced your family?**
   My two children were born while I was a tenure-track faculty member at a regional university. While I cannot say for sure that having a family influenced my career, I am certain that my career has influenced my family. My two daughters have only and always known their mother to work. This meant that we hired college students and other caregivers to be with them. Instead of feeling guilty about this, I tried to celebrate the different gifts that each caregiver brought into their lives: one sang to them when they were little, and hence, both girls can carry a tune (while I cannot). Two of the college girls were great at bringing crafts, so the girls are able to create things, and yet the sight of a glue gun terrifies me.
2. **Did you have role models? Which examples were set for you in your childhood or while you were growing up?**
   I was born in 1959 and I did not know very many women of my mother's age who worked. My own mother taught school for a few years before I was born. Anne Plyer was the mother of one of my friends: her husband was a stamp-collecting friend of my father's and she was a chemist who worked with my dad. She was the only professional woman I knew growing up.

3. **Have you come up against any significant obstacles during your career and how did you overcome these?**
   I would not say that I have encountered any significant obstacles in my career.
4. **Is there anything you would have done differently or would not do again?**
   Women of my generation, who went to college in the late 1970s, had many choices available to them, but not nearly as many as today's young women. I was not encouraged to aim high (to attend a great school, for example) and that perhaps limited my career. But I have no regrets.
5. **What advice would you give to young women hoping to pursue a career in academia? E.g., while studying, when planning a family**
   Many of today's young women are more aware of their options than I was at their age. They have heard about the fertility struggles of women who postponed childbearing into their 30s and beyond. My only piece of advice would be: always do your best (then you have no regrets) and don't worry about what other people think of your choices (they are too busy worrying about their own business). Both of my daughters have expressed an interest in becoming faculty members and I am glad that my experiences have not steered them away from this possibility. One is about to graduate with a degree in psychology and the other is a freshman biochemistry major—perhaps there will be more chemists **and** professors in our family.

My Mom and Science

By Alison Frech, age 21

Thanks to my mom, I wasn't privy to the fact that women hadn't always been encouraged to pursue a career in academia until I was nearly in middle school. I had long assumed that every mother was as delighted and involved with her career as my mother. This was enhanced by the fact that my parents conducted an annual chemistry demonstration at my elementary school, which elevated me to a celebrity status among my peers during the week they were scheduled to perform their science show. Together my mom and dad would dazzle us with beakers spurting copious amounts of foam or concoct liquid nitrogen ice cream. My mom was no assistant; the two of them ran the demonstration garbed in crisp lab coats, working as a team to interest the students before them in chemistry. This is illustrative of how I viewed their careers. Although I later learned about their separate focuses within chemistry, the perception that my mom was equally as capable as my dad has remained constant. Having such a strong role model in my life has allowed me to grow up sure of the fact that I am able to pursue and excel any career I desire, for my success relies on my ability, not my gender.

Alison, Cheryl, and Emily Frech on a return visit to Scotland in 2011.

Main Steps in Cheryl's Career

Education and Professional Career

1981	B.S. Biochemistry, Oklahoma State University, OK
1984	M.S Analytical Chemistry, University of Oklahoma, OK
1985	Visiting Research Assistant, Technical University Berlin, Germany
1987	Ph.D. Analytical Chemistry, University of Oklahoma, OK
1987–1988	Postdoctoral Fellow, University of Oklahoma, OK
1988–1989	Postdoctoral Fellow, Max-Planck-Institute for Polymer Research, Mainz, Germany
1989–1991	Visiting Assistant Professor, University of Oklahoma, OK

1991–1997	Assistant Professor, University of Central Oklahoma, OK
1997–2001	Associate Professor, University of Central Oklahoma, OK
2000–2001	Visiting Academic, University of St Andrews, UK
2001–present	Professor, University of Central Oklahoma, OK

Cheryl is a fellow of the American Chemical Society and an Associate Editor and an Editorial Advisory Board member for the Journal of Chemical Education. In 1999 she received the UCO Neely Award for Excellence in Teaching.

Upward Bound to a Ph.D. in Chemistry and Beyond

Judith Iriarte-Gross

"Mom! I am the only kid in high school who knows what an NMR is!" My career path to a Ph.D. in chemistry included raising a son...as a single parent. Once I made the decision to go back to college and to earn a degree, I realized that a career and a family life were hard to balance, but it could be done. I wanted a better life for my family and realized that this included getting a college education. This path to a Ph.D. in chemistry started in a small way in high school though I must admit that I did not like high school chemistry. I can say that I embraced a career in chemistry after working with some amazing mentors who helped me navigate undergraduate and graduate school as a student and as a single mom.

J. Iriarte-Gross (✉)
Department of Chemistry, Women In STEM Center, Middle Tennessee State University, Murfreesboro, TN 37132, USA
e-mail: Judith.iriarte-gross@mtsu.edu

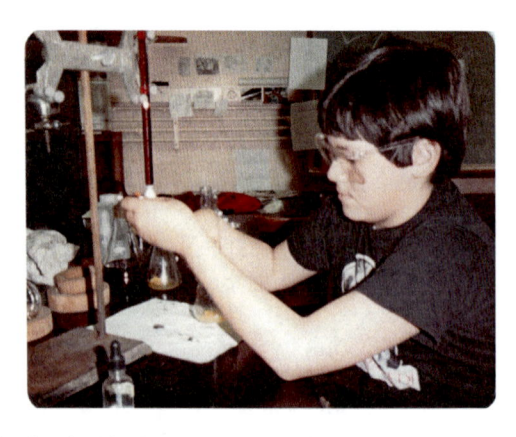

Daniel practicing titration for his sixth grade science fair project

I grew up in Prince Georges County Maryland, a suburb of S.E. Washington D.C. I am the oldest of seven children and one of two girls. My mother was always a stay-at-home mom and did not attend college. She did not know how to drive a car and still does not drive. My father divorced my mother when I was 13. Imagine seven kids, no transportation, very little child support, and no job. After looking back at this time in my life, I can honestly say that the Daughters of Charity at the Catholic schools that I attended, from kindergarten to grade 12, were my first mentors. They made sure that we received food baskets, toys, and clothes at holidays. I was given a scholarship to attend the girls' Catholic high school and bus money to make sure that I got to school. One year I was asked to work in the high school chemistry lab washing glassware and cleaning up the benches. Little did I know that my first introduction to chemistry would start me on a path to a Ph. D. in chemistry. In fact, at that time, I did not appreciate the beauty of chemistry but did appreciate the beauty of clean glassware! I did not realize that my high school chemistry teacher, Sister Mary Ellen, saw that I was an apprentice chemist. Today I tell my students that mentors have X-ray vision and can see things about you that you cannot see yourself.

During high school, I worked as a paper "boy" for a Washington daily newspaper. Keep in mind that during the late sixties, there were strictly defined job titles for males and females. I also babysat children in the neighborhood, a traditional job for a teenage girl. As a junior and senior in high school, I worked at Ford's Theater and National Theater in the evenings and on weekends as an usher. It was nice to have some spending money! I was a good student in high school but was not thinking about college. I did not know what I wanted to be when I "grew up." I knew that I did enjoy science. If I had any thoughts about college, I did not have a clue about how to start the college application process or where I would find the money to pay for college tuition. No one in my family had ever attended college or knew how to apply and I was told that there was no money for college. One day, however, the Sisters spoke to my mother about a new program for low-income students who had the potential to go to college. This program even paid the students a stipend for transportation, $10.00 per week! This program was called Upward Bound and was held at Trinity College in Washington D.C. Once again, my mentors provided me with a door to a better future, an opportunity for a college education. I tell my students today that one should always listen to a mentor because his or her advice can take you farther than a car!

I attended Upward Bound at Trinity College for three years. I was delighted to be able to live on the campus during the summer and to not have to watch over my younger siblings. I looked forward to the Saturdays during the school year, where I could take the bus to Trinity College for classes, tutoring, lunch, and $10.00. Upward Bound showed me the possibilities that a college education could provide to a first-generation student who was eager to learn. Upward Bound showed me that I could succeed in college by providing me with supplemental instruction, tutors, and opportunities to explore majors. During the summers, I lived on campus, ate in the cafeteria, and attended classes. One year, I asked for a science and math tutor and was able to work one-on one with a Trinity student who was hired just for me! In addition to academic classes, we learned how to read college catalogs and to complete applications for admission and for financial aid. We took practice SAT exams and learned how to improve our study skills. Each summer, Upward Bound took us on trips to visit colleges and universities. We also took cultural trips to concerts and shows. I remember seeing the Alvin Alley Dance Theater and hearing Ray Charles at a concert. We also attended a Washington Senators baseball game. Life was more than watching my brothers and sister and I began to imagine the possibility of attending college as a science major.

At home, I was still expected to watch my brothers and sister and this turned out to be a major roadblock to college. So in a way, my family: brothers, sister, and mother did influence my career. I had decided that I wanted to major in a science and had applied to several colleges on the east coast, hoping for scholarships and grants. I found out later that I had received acceptance letters, but my mother did not understand that financial aid offers were sent in separate letters. She turned down offers of admission without telling me because there was no mention of funding. I learned about this several years later. I was accepted to the University of Maryland, a bus ride with two transfers, away. I could still live at home and attend college. I could still live at home and babysit. This was not my plan. I desperately wanted to get out of the house so I married the man I was dating, two months after I graduated from high school. I could attend college and live in an apartment and not have to babysit! At the time, this was heaven. But soon afterwards, I discovered that this was a mistake. I was pregnant by the end of the fall semester. I dropped out of Maryland, had my son, Daniel, and started a job as a clerical worker at an insurance company in Washington D.C. My college career was over, so it seemed.

Reviewing insurance applications and ordering new id cards for clients was not the life that I imagined, but we did get health insurance, the office was on a bus route, and my salary helped to pay the bills. After two years of marriage, I realized that my marriage was a huge mistake and moved out with my infant son. One year later, my ex-husband was killed drunk driving. My job at the insurance company paid the bills and provided health insurance; however, I wanted more for myself and my son. I learned to drive for the first time and bought a car. I returned to Prince Georges Community College as a part-time student with hopes of becoming a pre-med major.

My first class was held in the evening at Andrews Air Force Base. It was a college algebra class with no more than two or three women in the section. Keep in mind that I was a clerical worker for five years. Clerical workers do not use algebra. I knew that I could ask questions so when the teacher started discussing "slope," I raised my hand and asked him to define a slope. He drew a "slope" on the board and laughed along with the class. I was not happy, but I was very determined. I told the teacher after class that I asked a legitimate question and that I would earn an "A" in his class and I did. I tell my students this story and remind them that no question is a silly question and that my job as a professor and mentor is to facilitate their learning.

This class and the few other classes that I took as a part-time student gave me the courage and the time to do the unthinkable: resign my job and go back to college as a full-time student. I needed to know that I could succeed in college. I also needed to know that I could afford college as a single mom. I was terrified of quitting a secure job, but I had to quit in order to move forward in my life and to find a satisfying career. It helped that my son was now in kindergarten and in a Head Start program which reduced my babysitting costs. My Upward Bound training kicked in and I was accepted as a full-time student at Prince Georges Community College, Maryland. I knew how to apply for financial aid and to search for private scholarships and grants. From the federal government, I received a Basic Education Opportunity Grant, BEOG, better known as the Pell Grant of today. I also wrote to my state senator, Steny Hoyer, for assistance with finding financial aid for college. Remember that there is no such thing as a silly question! As a result of my question to then state Senator Hoyer, I received a two-year full scholarship from the state of Maryland. Today, Congressman Hoyer represents the fifth District of Maryland and is the second-ranking member of the House Democratic Leadership. Though I have never met him in person, I am truly grateful that he took the time to provide a constituent with the help necessary to raise a child and earn a college degree. I consider him a mentor who supported my decision to return to college full-time at a time in my life when I had many doubts.

Now that my son and I were both full-time students, we settled into a routine. Wake up, pack lunch, eat breakfast, and head to school. After school, relax a little, prepare dinner, do homework, and get ready for bed. During the weekend, we cleaned house, washed clothes, studied, and visited the Smithsonian. I was a college student, pre-med major, and a mom. Life was hard, but finally, life was meaningful.

I enrolled in my first general chemistry course in the 1978 fall semester. I relearned the language of chemistry and fell in love with dimensional analysis. Where was dimensional analysis when I needed it in my math classes? In the spring 1979 semester, I truly enjoyed thermodynamics, balancing oxidation–reduction reactions, and solving chemical equilibrium problems in gen chem II. I still have my gen chem textbook, Nebergall fifth edition, though minus its front cover. My professor, Dr. Pat Cunniff, encouraged me to major in chemistry and suggested that I attend a summer program for women and minority students at the University of

Maryland professional school in Baltimore. I attend that program and she was right. I was not a pre-med student but a chemistry major. I listened to my mentor!

I flew through classes at Prince Georges Community College and transferred to the University of Maryland as a chemistry major. My son was growing up fast and was in elementary school full-time. A major obstacle for any parent is to find an amazing and trustworthy babysitter. I was fortunate that there was such a woman in the next apartment building who took care of neighborhood children like they were her own. How do I know this? Late one night, my son and I were outside watching something in the sky. It was too long ago and I don't recall the astronomical event that had us up late. It was dark and cold outside and our babysitter saw us standing in the parking lot. She hurried down thinking that her "baby" was hurt and we were waiting for an ambulance. Great babysitters are precious like gold and make a significant obstacle to work and family balance, childcare, turn into a small issue that is easily solved.

I mentioned earlier that there is no such thing as a silly question. I asked my way through my undergraduate program. Who is the best calculus teacher? I asked the chair of the math department and found an amazing calculus professor. How can I survive calculus-based physics? Visit the physics tutoring lab and practice problem solving. Should I take physical chemistry during the summer? Yes! I took both physical chem I and II during the summer session. I regret that Dr. Sandra Greer, 2004 Garvan-Olin Medal Recipient and winner of the 2014 ACS Award for Encouraging Women Into Careers in the Chemical Sciences, was not my physical chemistry professor. She is an outstanding advocate for women in chemistry and in STEM. Today, I consider her a mentor. What would undergraduate research do to help my career? Undergraduate research will introduce you to opportunities beyond that in a traditional laboratory course. I did find an undergraduate research mentor, Dr. Mike Bellama, who introduced me to basic research methods. My son was in sixth grade and I graduated with my B.S. in chemistry in 1981.

I am very happy to be the first in my family to graduate from college. I earned my B.S. in chemistry and shared my first graduation photo with Daniel

Life was good and I knew that I wanted more so I applied for and was accepted in the M.S. program at Maryland. Life was good but not good enough to go to another university out of state which would mean finding a new babysitter, new school for my son, and new housing. At Maryland, I was eligible for graduate student housing. My son was now ten years old and convinced me to give him a house key. The chemistry department was a 5 minute drive or 15 minute walk away from our apartment on the College Park campus. I was first nervous about leaving him early in the morning, 30 minutes before the school bus arrived, but several kids in graduate housing caught the bus and he did not want to be late in front of the other kids. He also checked in when he got back home with a quick phone call to my lab. Life continued to be good.

I conducted my M.S. research at the National Bureau of Standards, now called the National Institute of Standards and Technology. I synthesized barium and titanium sol-gel coatings and learned how to use an inert atmosphere glove box and how to tune an NMR. However, driving to Gaithersburg every day caused some stress, due to my son being at home in College Park. I mentioned earlier that babysitters are gold though we do not call a caregiver for a 10 year old, a babysitter! I found other moms in graduate housing who were happy to keep an eye on my son while I was off campus. Though I must say that 95% of these moms were not graduate students but were married to graduate students. As an undergraduate at Maryland, I had one class with one woman visiting professor and I did not have any classes with women faculty in the M.S. program. I was fortunate to find mentors at Maryland who understood that I was a grad student and a mom. However, I did start wondering about the absence of women faculty in chemistry and the sciences. I earned my M.S. in inorganic chemistry in 1984, and with a 12-year-old son, I was ready to start a Ph.D. program out of state.

Today I tell my undergraduates to ask many questions as they explore graduate program opportunities. What is your policy on comprehensive exams? Do you have to choose a research group by the end of your first semester? Can one interview both grad students and research directors before choosing a group? Is health insurance provided? How many seminars do you expect a grad student to give? Are there travel funds to present at professional meetings? What other training (proposal writing, IRB, safety) is provided to grad students? Will you pay my travel to visit your program and campus? Ask questions before you make that critical decision about a graduate program. I must admit that I did not ask all of the questions that I should have asked when considering Ph.D. programs. Ask questions that are important to you. I did mention that I was a single parent with a son. One professor told me about the preschool program on his campus. I quickly learned to mention that my son was in eighth grade and that I was not the typical grad student with younger children. After many questions and three campus visits, I decided to attend the University of South Carolina where I could afford to live on my graduate student stipend and raise a growing boy.

Our move to Columbia was not without problems. I discovered that I had to have knee surgery and postponed my Ph.D. program to January 1985. Once we arrived in town, my son was very nervous about starting a new school in a new state in the middle of the school year. He broke out in a rash and I had to find a local doctor during the first week that we were in town. He spent the first 12 years of his life in the Washington D.C. area and was scared to leave all of the family behind. He also thought that he would be called a "Yankee." He was worried about a class that he (and all students) was required to take in seventh grade in Maryland: a quarter semester of sewing, shop, cooking, and Spanish. He did not want to be teased by new friends in South Carolina for taking sewing and cooking! He could deal with shop and Spanish. He was worried about his math skills and if he had to read extra literature books to catch up. His worries were legitimate but turned out to be groundless. He quickly made friends and, to my knowledge, was never called a "Yankee" or even a "Damn Yankee." He was ahead in math and helped his new friends understand "The Hobbit" which according to my son was disliked by all. Remember that this was over twenty five years ago before the Hobbit films that we love today. We both agreed that the spring 1985 semester was a tough one without family and longtime friends in the neighborhood, but we survived. Thinking aback on this major change in our life, I would not have done anything differently. I took the time that I needed to be "Mom" apart from my being a graduate student. Make time for your child because he or she will grow up so quickly. Your research will still be there for you.

We settled into a routine once that spring semester was over. During the summer, my son hung out with his new friends and I hung out in the Odom research lab. I took my time to find an understanding research advisor who accepted that I was both a single mom and a graduate student. It was important that whomever I worked for would recognize that sometimes I had to be a mom, especially a mom to a teenage son. I decided to work for Jerry Odom for several reasons. I was intrigued by the vacuum line chemistry at NIST and thought that I would enjoy learning new techniques. I also wanted to move to p block of the periodic table. I wanted to learn new chemistry and found that Group 14 and Group 16 chemistry was the focus in the Odom research group. I could learn new synthetic methods using my personal vacuum line! The Odom group also used NMR extensively to characterize products and to follow reactions, but this was not the usual ^{1}H or ^{13}C NMR. We played with ^{29}Si and ^{73}Ge as well as ^{77}Se and ^{125}Te. What fun! My son and I were finally at home in South Carolina. Daniel was happy at Irmo High School, home of the world famous, so I was told, Okra Festival. I don't recall eating okra except if it was hidden in soup. I never ate the favorite southern breakfast food, grits, while living in South Carolina. This was our life in South Carolina!

My classes and comprehensive exams were over and my focus was on my research. I spent many long weekends running NMR experiments. My son would come with me and do his homework while I set up my experiments. He often commented that he was the only high school student in South Carolina who knew what an NMR experiment was! Afterwards we would go shopping, see a movie, or visit friends. This worked great until Daniel was 15, that age when children become very vocal about driving and wanting a car. However, I was a graduate student raising a growing boy on a graduate student's salary. "Mom, I need a car," was his favorite sentence though sometimes it changed to "Mom, I need a Mustang." He got his learner's permit, and during the weekends, I taught him how to drive on empty parking lots on campus. He learned to drive on interstates by driving me to campus, again on the weekends, when I needed to set up experiments. One morning he was trying to merge onto the highway and saw a line of tractor trailers quickly approaching his in lane. I laugh at this now because he was not asking for a car at this moment; instead he said over and over again "Trucks are coming!" After we arrived at campus and he realized that he survived highway driving, he still wanted a car!

My new saying became "Get a job if you want a car." He did and started working at a veterinary clinic on weekends and sometimes after school. We moved to South Carolina with one cat, Garfield, and one car. During our time in Columbia, our cat family grew to include Doris (named after Daniel's grandmother who does not like cats), Oreo, and Patches and a second car. There was peace at last for a while.

Time passed quickly and I was soon writing. Daniel was happy that I had reached this stage of my program and that my bench work was done. Raising a teenager while in graduate school was challenging and at times very funny. He did have a car but he also wanted a mom who did not smell like rotten garlic, the smell of selenium compounds. Each evening, I had to change clothes after a day in the lab because of the "garlic" (selenium) smell. We were in the grocery store one day and the clerk remarked that she smelled something funny. Of course, my son assumed that it was me and was totally embarrassed. He did not want to be seen in public with me in my fragrant lab clothes. He took over the grocery shopping and often made dinner during the week. I taught him how to use the best kitchen appliance in the world: the Crock Pot! He learned to make chili, soups, and became quite good with Crock Pot chicken and potatoes. We were a team with a common goal: graduation for both of us!

Dreams do come true. Daniel graduated from high school and I earned my Ph.D. in Inorganic Chemistry from the University of South Carolina

I completed my dissertation and was preparing to defend when a hiring freeze was put into place on campus. Jerry asked me to put off defending my dissertation till the very last day since he would not be able to pay me with the hiring freeze in place. See what I mean about finding the best research mentor! The defense day finally came and Daniel was in the audience. The presentation went smoothly and the questions began. I could see Daniel squirming in the back of the room. He did not understand the "defense" process and was getting upset about the nonstop questions. I reached a point where I could not answer a question and the defense was over. A lesson learned is to explain to your older children about the process that you are going through to earn an advanced degree. It requires long hours in the lab, in library, and writing. Grad students must be able to communicate their chemistry as a teaching assistant in the undergraduate lab, giving a departmental seminar, or presenting at a professional conference. Tell your children that being a grad student or research scientist is not the same as a 9–5 job but depends on the chemistry.

My son graduated from high school three days after I passed my defense and received my Ph.D. in inorganic chemistry. I would have never thought that one day I would earn a Ph.D. in chemistry when I was growing up in Washington D.C. Even today, I still am amazed that I have a Ph.D., but this should not be a feeling of amazement but one of confidence. This is another lesson learned that I share with my students. You can do anything if you have the determination and drive to succeed. Find mentors who understand that work (both undergraduate and graduate school) and family need to be balanced in a way that is best for you. Keep your mentors informed even when chemistry has to be pushed to the back Bunsen burner. Don't be afraid to ask questions. If you can't feel comfortable talking with your mentor then that partnership will not succeed. Remember the goal and do what is best for your family and for your science.

I moved to Dallas, Texas as a postdoctoral research associate in the lab of Patty Wisian-Neilson at Southern Methodist University. Continuing my tradition, I joined her group to learn new skills and polyphosphazene chemistry. I moved to Texas with four cats! Cats were not welcome in residence halls! Daniel stayed in South Carolina for college. He studied finance and now lives in Maryland with Bonnie, his wife, and their three children, Elaina, Waverly, and Logan.

I was still trying to find my chemical career path after my postdoc. I worked at the FDA lab in Dallas where I was an analytical chemist. It was an interesting job with great benefits and never boring. After a day identifying pesticides in cantaloupe or measuring how much mercury is in a shark, I was teaching intro chemistry at a community college twice a week. I really enjoyed interacting with students more than interacting with a GC-MS. I was not an analytical chemist, so I found a job in Fort Worth as a synthetic sol-gel chemist and lab manager. I soon discovered that a job in industry was not for me. After a search for a tenure-track position, I joined the faculty of the Department of Chemistry at Middle Tennessee State University in 1996. I am delighted to say that my husband, Charles Gross, a native Texan, retired and followed me to Tennessee. His job is now to take care of me, our cats, and our home.

Today I am a full professor and Director of the Women In STEM Center on campus. I am an advocate for our women students who sometimes still question if they can have a career in chemistry and a family. We share stories, discuss options, and offer possible solutions. I encourage my students to consider all opportunities such as undergraduate research, internships, and professional presentations which will enhance their resume. I ask women students to step outside of their comfort zone and take on leadership roles on campus. Women can juggle and balance both a career in chemistry and a family. I know because I have a wonderful career in academia and a son who does know an NMR!

Daniel did not receive a car for his high school graduation. He did receive this t-shirt!

Main Steps in Judith's Career

Judith Iriarte-Gross earned her B.S. and M.S. in Chemistry from the University of Maryland, College Park, and her Ph.D. in Inorganic Chemistry from the University of South Carolina. She completed a postdoctoral research project at Southern Methodist University. Before joining Middle Tennessee State University (MTSU) in 1996, she worked as a chemist for the FDA and as a chemist and lab manager in the plastics industry. Dr. Iriarte-Gross mentors an active undergraduate research group in the scholarship of science education and has been involved with SENCER, Science Education for New Civic Engagement and Responsibilities since 2005. Dr. Iriarte-Gross is nationally known for her advocacy for encouraging girls and women in the sciences. She was named an Association for Women in Science (AWIS) Fellow in 2009. In 2009, she was named the director of the Women in STEM Center at MTSU, the only center for women in STEM in Tennessee. She is currently co-President of the Tennessee Chapter of AWIS and represents AWIS on the National Champions Board of the National Girls Collaborative Project.

The Window of Opportunity

Nancy E. Levinger

Photo: Peter B. Seel

As a beginning graduate student, well before I even met the man I would eventually marry, I recall considering my future. I was training to become a scientist, but would I find a partner? Would my career path allow me to have a family? Although these basic personal choices had always seemed inevitable to me as a child, in the frenetic schedule of a chemical physics graduate student they were anything but given. At that point, studies showing the impact of advanced education on women's personal lives had yet to appear [1]. Still, it seemed clear; the likelihood of finding a partner while spending almost all my waking hours working on science was probably pretty small. After many musings, I made an active decision: the rich career afforded by my advanced degree would fulfill me whether I married or not, or had a family. This decision played a role in my success as a graduate student. It allowed me to focus my energy on science and leave my personal life to chance. I thrived in graduate school, both academically and personally. About two years later, I met Pete, the man who would become my best friend, my husband, the father of my children, and the person who made it possible for me to succeed in my career as a professor and as a mother.

So what does success look like for Prof. Mother? What kinds of obstacles existed for me as a graduate student in the 1980s, an assistant professor in the 1990s, and now as a full professor [2, 3]? What extra challenges arose because of my personal

N.E. Levinger (✉)
Department of Chemistry, Colorado State University, Fort Collins, CO, USA
e-mail: levinger@lamar.colostate.edu

choice to raise a family? Although I do not claim to have universal answers to these questions, I believe that my life provides insight into the things that make it possible for women to thrive simultaneously as academicians and mothers, and some of the significant issues that still remain. The individual path for each woman will vary but some well-considered choices can increase the chances of reaching satisfaction in both career and personal life.

Early Career Decisions

I believe that the decision most critical to successfully balancing career and family rests on the choice of partner. An academic career places tremendous demand on individuals regardless of their personal choices. A supportive partner understands the career demands. A family also places demands on parents. A supportive partner understands and happily steps up to share the demands that come with raising children.

When Pete and I decided to marry, we discussed whether we would have a family. Both of us hoped that our union would include kids. Our wedding vows included a line "I take you…to be the mother/father of my children" (which overjoyed my own father who until that point was not sure we wanted to have our own children). When we would actually fit this into our research-full lives remained a mystery. Finishing my Ph.D. and moving on to a very demanding postdoctoral fellowship, I wondered if we would find a way to start a family. Already in my 30s, we knew that risks associated with pregnancy would only increase and that our energy to keep up with kids would wane as we grew older. When my search for an academic job loomed in the summer of 1991, Pete and I decided that we had a window of opportunity. We would stop using birth control for four months from June to September; if I became pregnant, we would have a baby in between my postdoc and professor jobs. If not, we would wait. Three months passed without a positive pregnancy test, but in September 1991, the last month of our window of opportunity, I became pregnant. We were elated and nervous. This meant that my academic job search would include balancing preparation and, hopefully, interviews while pregnant.

In 1991, just being a woman seeking a faculty position at a research-intensive institution placed me in a small minority. Adding pregnancy to that mix was not something I wanted to broadcast, so, except from a very small number of close friends and family, I hid my pregnancy. I applied to jobs posted at 21 different institutions and was thrilled to receive invitations for several interviews. During January 1992, I interviewed at three different institutions, which was also the beginning of the second trimester of my pregnancy. Though I had already gained weight, I felt that the pregnancy was not yet really showing. However, that did not stop a potential colleague at one of the interviews from stopping dinner conversation to ask me (in front of four other potential colleagues), "So, Nancy, it looks like you have gained some weight since we last met. Do you want to tell us why?" I was mortified and answered, "No." Years later the professors present at that fateful dinner told me that they knew I was pregnant and it did not matter. Whether it mattered or not can never be known. Nonetheless, I feared that if I confirmed the pregnancy it might jeopardize my possibility to receive an offer from that institution or anywhere else. When I finally revealed my pregnancy to my postdoctoral advisor, he was shocked but supportive. In the end, I was thrilled to receive an offer to join the faculty at Colorado State University. I might have received other offers but having grown up in Fort Collins, CO, and with my parents and sibling still living there, this opportunity seemed like a dream come true. My Colorado State University colleagues speculated that I was pregnant, a fact I did not confirm until I had accepted the job offered to me.

My first child, Ian, was born early in July 1992. Six weeks later, I started as a brand new assistant professor of chemistry at Colorado State University. Four years and two miscarriages later, my second son, Eric, was born in late July 1996. Both births and all pregnancies occurred prior to my earning tenure. In 1995, Colorado State University began considering options for probationary faculty to extend the tenure clock for personal reasons such as childbirth. With a newly passed policy [4], I may have been the first faculty member at CSU to extend the tenure clock on the basis of childbirth. Exercising the option to delay my tenure decision reduced stress and gave me time to fill gaps in my academic vita. The extra year of uncertainty about my ability to meet the expectations and earn tenure added to personal stress. Through three pregnancies (two miscarriages and one viable) I had spent ~15 months of my time as an assistant professor pregnant. Having balanced work and caring for infants my curriculum vitae had obvious gaps. From frank discussions with more senior colleagues, I know that some questioned my productivity (or lack thereof) during the time of pregnancy and infant care. Although data suggest that extending the tenure clock can lead to salary inequities, the data also show positive impact on promotions [5]. I continue to be grateful for the opportunity to postpone the tenure decision, an opportunity that is now nearly universal at academic institutions for women and men alike.

Not knowing the pressures presented by the simultaneous start of my independent academic career and family was probably good for me. Had I understood the challenges that both would place on me, I may never have chosen to have children. As it was, we managed. With the help of two spectacular nannies, I learned to balance demands of a career at a research-intensive university with the needs of infants, toddlers, and young children. Initially, my husband worked in Boulder and then in Denver, more than one hour's drive away so child emergencies were entirely my dominion and there were plenty of them—the nanny calling in sick at 7 am when I had to teach my class at 9 am (Grandma to the rescue!), the fall that four-month-old Ian took yielding a goose-egg sized lump on his forehead and a skull X-ray (he was fine). A few months before Eric was born, my husband achieved his career goal of gaining employment at the Hewlett-Packard site in Fort Collins. Trading his hour plus commute to work with a 10 minute by car, or 30 minute by bicycle, commute drastically improved our quality of life. He was much less tired and so much happier, which made him more able to support my frenetic schedule.

Eric, Nancy, and Ian while on sabbatical leave in 2000

Academic Success

When I entered graduate school after college, I was pretty sure that I wanted to pursue an academic career. The balance of research and teaching seemed like something I would really enjoy. My experiences as a teaching assistant assured me that I liked teaching; research, although demanding, was also stimulating and exciting. Shortly before earning the Ph.D., I discussed careers with my graduate advisor. When I expressed interest in an academic career, he probed me, asking what kind of academic job I felt would be best. At that point, I could not imagine what kind of research I could do at a primarily undergraduate institution so I aimed for a position at a research-oriented university. In hindsight, this seems like a rather random way to pursue this demanding career.

During my postdoc, a workshop led by Dr. Tom Blackburn, then a program officer from the ACS Petroleum Research Fund, provided critical guidance and gave me confidence that I could succeed in as a professor at a research-intensive university. Tom, assisted by University of Minnesota Professors Peter Carr and Larry Miller, led a group of chemistry postdocs and graduate students through a short exercise in which we came up with ideas for research proposals. Even though I had garnered a prestigious NSF Postdoctoral Fellowship, I doubted my ability to generate fundable research ideas. This short exercise at a critical junction in my life demonstrated to me that I had lots of fundable ideas. The confidence boost was enough to encourage me to apply for faculty jobs at research-intensive institutions. You might expect that gaining this confidence would be enough to allay my self-doubt from then on, but lagging confidence would continue far into my career.

As an assistant professor, I applied broadly to granting agencies for funding. My applications yielded early fruit, netting funding from the ACS PRF and then a prestigious NSF Young Investigator award. I also teamed up with colleagues to write two instrumentation proposals, both of which received funding. Although money came rather easily, papers were much more challenging. When my first full paper was rejected for publication, I did not know what to do. I had never experienced this as a graduate student or postdoc. Those results remain unpublished to this day because I did not realize that one could simply revise the paper and resubmit. Now I serve as a mentor for junior faculty, and in this role, I hope to preclude some of the mistakes I made.

Role Models and Encouragement

As society started to accept women's abilities, most women of my generation, born towards the end of the "baby boom," did not encounter the enormous overt barriers to pursuing science faced by earlier generations ([2]; [6]). Without doubt, my father had the most significant influence on my early scientific interest and success. I remember the excitement and fascination I had when my third grade class had a unit studying astronomy. Realizing my interest, my dad encouraged me to explore much further than my third-grade class. He pulled a book of star maps from the shelf and he and I poured over them to figure out what we could see in the dark night sky. Together we marveled at science in the first episodes of NOVA that began airing on public television around 1973. I was particularly enamored by the episode entitled "The First Signs of Washoe," reporting about a chimpanzee who learned sign language. By the time I started junior high school, I sought and received the opportunity to take science instead of the required home economics course. I also participated in a program entitled, SCIP (Science Careers Investigation Program) that took girls and underrepresented minority students out of school for field trips to encounter science firsthand. By the time it came to apply to college, I knew I would pursue a science career.

For most of my career, even though I had wonderful academic mentors, I did not have a female role model. In retrospect, I realize that a few role models existed, but I did not connect with them. My rejection of female role models puzzles me now but fits a well-documented pattern. Raised in the same society, women are just as likely to demonstrate implicit bias toward men in male-dominated fields and roles [2, 7, 8]. By now in 2014, most overt gender discrimination is a thing of the past. Unfortunately, covert gender bias is still alive and well. We all have biases. We use many of them automatically to make the decisions we constantly face in life. The problem arises when bias limits our ability to pursue or achieve our goals. As an assistant professor, I received a prestigious NSF Young Investigator Award (the predecessor to CAREER). A male colleague of mine, two years ahead of me on the tenure track, had also applied for the award but did not receive it. Rather than congratulate me on my award, he told me that I had only received the award because I was a woman. Needless to say, his comment fed my insecurity making me doubt whether I really deserved the award. Although it should be a thing of the past, 20 years later a young female colleague of mine suffered the same response from our young male colleague when she received a prestigious award.

In my current position, I continue to struggle for gender equity. As a young professor and parent, I was told by the chair of our promotion and tenure committee, "Nancy, your priorities are not right. You need to place research way above everything else, significantly above your familial obligations and way above teaching and service." I responded that family and research could be equal but I could not place research above my familial obligations. I hope that no junior faculty member of mine would receive the same demand. My challenge now is to recognize barriers when they arise. This may sound odd—we should recognize a barrier in the way of our progress. But often the discrimination can be difficult to identify. This is particularly true for the standards to which women are held compared to men.

Impact of Career on Family and Family on Career

We usually think of family influencing career, but career can also influence family. Sometimes it is hard to figure out which way the arrow points, Family $\rightarrow$ Career or Family $\leftarrow$ Career? Surely those temporary issues like dealing with a sick child or having to pick kids up from day care on time fall under the category of family impacting career. Having meetings to attend out of town or work obligations that must get done fall under the category of career impacting family But most of the time, we rely on our understanding of chemistry to understand the interaction. We seek balance in equilibrium Family $\rightleftharpoons$ Career. So how have I found this elusive equilibrium?

For the first eight years of our children's lives, my husband and I chose to hire a full-time nanny, the most expensive but also the most convenient childcare. Initially, my husband's entire take-home pay barely covered the nanny's salary and the house mortgage, but it was worth every penny that we spent. We were blessed to have two exceptional nannies Randi and Donna for all but four months of those eight years; in between these two, we had another nanny who did not work well with us. Our nannies did not live with us. They arrived at our home at about 7 am and stayed until 6 pm. They did so much more than just tend to the children—all the laundry, much of the shopping, some food preparation—in addition to providing an attention-rich environment for the boys. After both children were in school full-time, we hired a series of wonderful after-school babysitters who picked the kids up and cared for them until we got home. Each had her own style and the kids loved them all. We continue to maintain contact with all of these wonderful women many years after they stopped working for us. Our attention to our nannies' needs helped them to be able to stay with us for years. Indeed, we have continued to help both in times of need and they have returned favors for us. Most of the weddings our kids have attended were of their former babysitters!

One way that my career has influenced our children is through their exposure to and reliance on lots of people other than their parents. Ian developed lasting ties to

Randi, who cared for him from age 6 weeks to more than 4 years. Eric developed significant ties to Donna, who began caring for the children when he was ~8 months and only stopped caring for them when we left for our first sabbatical leave, a few weeks before Eric's fourth birthday. Both nannies worked as a team with my husband and me. They echoed our values, read to our sons, took them to enriching activities like swimming lessons, and more. Part of who my sons are today comes from their strong relationships with these wonderful nannies.

When the children were young, I never felt as though I was doing enough, neither as a faculty member nor as a mother. I remember discussing this with a friend when Ian was about four years old. I intimated my concern that I was not spending enough time either at work or at home. Ian piped up and said, "but Mama, you spend lots of time with me." At that point, I knew that even if I did not spend all my time with him like many of my stay-at-home-mom friends, I spent *enough* time with him. I tried to stop berating myself about the amount of time I spent with my family and focused on making that time the best time possible.

My constant interaction with college students has affected the way that I treat my children. As a professor, you would think that the "the dog ate my homework" stories would stop and that you would not be subject to the interference by the parents of our college students. Seeing students perform below their ability in college has colored my view of my own kids' futures. Dealing with angry, accusatory parents of college students has impacted the way I treat my sons. In cases where other parents intervene and advocate, I am much more likely to take a backseat and expect my child to solve his problem himself. Perhaps this put my kids at a disadvantage during elementary, middle, and high school. However, I believe that my expectation that they figure out solutions to their own problems will have long-lasting positive impact on my children. I hope that my expectations will lead them to take responsibility and initiative leading them to productive lives.

When asked how my career choice has impacted them, my sons responded predictably. First, they have never experienced life with their mother staying at home. Even if other mothers stayed at home, they did not feel that my career choice negatively impacted their lives. My older son noted that with a Prof. Mother, a child is never on vacation. When they would pose a question, invariably they would stimulate the Socratic method in their mother leading to many more questions than answers. Alternatively, a significant estimation would occur, like trying to figure out if there is a mole of grains of sand on Earth. My career choice has enriched my sons' lives, especially through sabbatical leaves taken away from home. Living in the Bay Area on three different occasions provided new experiences for the whole family.

Ian, Nancy, and Eric on their way from the Exploratorium in San Francisco, 2010

Nancy's 50th birthday in Barcelona with Pete

Advice and Recommendations

Frequently throughout my career, I have entertained the question, "How do you do it? How do you balance career and family?" Early in my career, this question was hard to answer. "You just do it." is probably the best I could manage in the beginning. At least once early in my academic career, I remember a woman graduate student who was taking the course I taught saying to me, "I don't want to be like you. I want to have time for my family!" This comment felt incredibly depressing. Instead of serving as a role model, I felt like an anti-role model. Indeed, balancing an academic career with spouse and kids has a different meaning for women academics than it does for men [9]. But this comment resonated with me and motivated me to speak about my experiences.

Advice is a dangerous thing to give and to take. Personal and professional situations vary so there is no "one size fits all." Still, I believe there are important lessons one can take from my career. Here are a few:

- Your choice of a partner is probably the most important variable you can control. Choose a partner who understands the demands of your academic career and wants to help you to succeed in it. Studies show that women tend to carry more than 50% of domestic responsibilities [10]. Choosing a partner who takes on significant domestic responsibility (cleaning, cooking, shopping, childcare, etc.) makes it possible to balance academic and personal tasks. Choosing a partner who is patient, supportive, and committed is the most important thing you can do to succeed in your position.
- When you are emotionally ready to start a family, stop using birth control. When you get pregnant, you will figure out how to work this into the equation. There are, of course, somewhat better and worse times to start or add to a family. But there is no real "right time" so waiting to start can lead to problems associated with being pregnant later in life.
- Find out what policies exist that can help you achieve your professional goals and exercise them if necessary. Policies to stop the tenure clock can seem dangerous so we must continue to educate colleagues that adding time to the probationary pre-tenure period should not raise expectations for productivity. Although slow to come, this understanding seems to be taking root. If you need to stop the tenure clock, do it and don't worry!
- If you are planning a family or expecting a baby, have a well-devised and comprehensive emergency plan. If your regular childcare falls through, have a backup plan. Who will you call? Who can help you? There are many people who want to help—let them!
- As soon as you have enough money, pay others to do the things you don't like to do or don't have time to do. Time is in short supply when you have a career, let alone career and family. It is worth spending money for someone to clean your house, do yard work, cook your meals, or whatever you would prefer not to do.

- Look for role models and mentors. Listen to their advice and work to incorporate it into your life.
- In his book, *The Four Agreements* [11], Don Miguel Ruiz lists four agreements to live by: (1) Be impeccable with your word (don't gossip); (2) Don't take things personally; (3) Don't assume anything; and (4) Always do your best. Of these, the first and last are pretty straightforward. Most of us do them without trying. We often interpret the words and actions of other people to mean that we are somehow wrong or inferior when in actuality, these words and actions reflect issues in the person serving them. Not taking these actions or words personally allows us to analyze the situation without becoming hurt or glamorized. Likewise, assuming that others understand us can lead to significant problems. Much better to remove doubt about your words, actions, and intentions. These agreements can be hard to follow, but if you can, your life will be easier.

Finally, find time for yourself. Life is short and you do not know where it will lead. Carpe diem!

Main Steps in Nancy's Career

Education and Professional Career

1983	B.A. Integrated Science and Physics, Northwestern University, IL
1990	Ph.D. Chemical Physics, University of Colorado –Boulder, CO
1990–1992	NSF Postdoctoral Fellow, University of Minnesota, MN
1992–1999	Assistant Professor, Colorado State University, CO
1999–2005	Associate Professor, Colorado State University, CO
2000–2001	Visiting Scholar, Stanford University, CA
2005–present	Professor, Colorado State University, CO

Honors & Awards (selected)

2005	Fellow of the American Physical Society
2004	Margaret Hazaleus Award, Women's Caucus, Colorado State University
1994–1999	National Science Foundation Investigator Award

Nancy is the founder of the NSF Chemistry REU Leadership Group. She is also University Distinguished Teaching Scholar and holds a courtesy appointment as a professor of Electrical and Computer Engineering at Colorado State University.

References

1. Mason MA, Goulden M (2004) Marriage and baby blues: redefining gender equity in the academy. Ann Am Acad Polit Soc Sci 596:86–103
2. Valian V (1999) Why so slow? The advancement of women. The MIT Press, Cambridge, MA
3. Minerick AR, Washburn MH, Young VL (2009) Mothers on the tenure track: what engineering and technology faculty still confront. Eng Stud 1:217–235
4. Colorado State University (2012-2013) Academic faculty and administrative professional manual, section E.10.4.1.2 Extension of the probationary period. http://www.facultycouncil.colostate.edu/files/manual/sectione.htm#E.10.4.1.2. Accessed January 7, 2014
5. Manchester CF, Leslie LM, Kramer A (2013) Is the clock still ticking? An evaluation of the consequences of stopping the tenure clock. ILR Rev 66:3–31
6. Monroe K, Ozyurt S, Wrigley T, Alexander A (2008) Gender equality in academia: bad news from the trenches, and some possible solutions. Perspect Polit 6:215–233
7. Latu IM, Stewart TL, Myers AC, Lisco CG, Estes SB (2011) What we "say" and what we "think" about female managers: explicit versus implicit associations of women with success. Psychol Women Quart 35:252–266
8. Committee on Science, Engineering, and Public Policy (COSEPUP) (2007) Beyond Bias and Barriers: Fulfilling the Potential of Women in Academic Science and Engineering. National Academy of Sciences, National Academy of Engineering, and Institute of Medicine. National Academies Press, Washington, US
9. Park B, Smith JA, Correll J (2010) The persistence of implicit behavioral associations for moms and dads. J Exp Soc Pyschol 46:809–815
10. Misra J, Lundquist JH, Templer A (2012) Gender, work time, and care responsibilities among faculty. Sociol Forum 27:300–323
11. Ruiz DM (1997) The four agreements: a practical guide to personal freedom. Amber-Allen, San Rafael, CA

Wanting It All ...

Cecilia H. Marzabadi

My Story

Growing up in the 1960s and 1970s when so many changes were happening in American society made me believe that my dreams could be fulfilled and that being a woman was no longer a barrier to the career that I wanted. I could be married and have children and have a successful career. However, along with "wanting it all" is the reality that you have to prioritize and learn to balance what is most important to you in life. This is my story of how I have managed to wear the many hats that I do: as a chemistry professor at a mid-sized, private Catholic university, as a research mentor, as a wife, daughter, sister, and friend, and as a mother.

As a child...I don't think I even knew any scientists. My mom was a commercial artist who drew department store ads for the newspaper. My dad was the owner of an industrial laundry. I am the only child from a second marriage; I have 2 half-brothers and 2 half-sisters. Growing up I had an interest in everything, including science. I loved reading and learning.

C.H. Marzabadi (✉)
Department of Chemistry & Biochemistry, Seton Hall University, 400 South Orange Ave, South Orange, NJ 07079, USA
e-mail: cecilia.marzabadi@shu.edu

When I began my undergraduate studies at Saint Louis University (SLU), I had no idea what I wanted to major in. I knew I wanted to work with people and that I enjoyed solving problems. I also enjoyed science in general. I took one of those career aptitude tests and it said that good careers for me based on my interests would be as a physician or as a college science professor. I took both biology and chemistry courses initially and after about a year decided that being a chemistry major was more suited to me. Biology seemed like too much memorization, whereas chemistry involved more problem solving. Organic chemistry in particular appealed to me; it wasn't as mathematical and had a large mechanistic component.

As an undergraduate I began doing research in organic chemistry in the area of organometallic chemistry. This enabled me to get to know the professors and other undergraduate and graduate chemistry students. It also gave me the feel of what it was like working on a research project in a lab; I loved it! During this time, I met my future husband, Mohammad, who was my teaching assistant in first semester organic lab. In my senior year, we began dating. Our relationship would significantly alter the subsequent career paths for both of us. He decided to stay local and pursue his Ph.D. degree at Washington University. I decided to stay at SLU and to get my research MS degree in organic chemistry. I had the hope that by the time I finished my MS degree, he would be finished with his dissertation work and we could plan our next moves together, including my further pursuit of a Ph.D. Well, as they say… "the best laid plans". Unfortunately, Ph.D. research doesn't always follow a time clock, so I finished my studies before he did and looked for a job in the St. Louis area.

I was hired by Monsanto Agricultural Products Company, working in their metabolism group. Although my background was in synthesis, I readily learned how to do the analytical work required in metabolism. It was a very interesting job that taught me many new skills, including working with biological samples. In spite of this, I knew that eventually I would be limited in advancement in industry without a Ph.D. degree. After 2 years at Monsanto, I left to pursue my doctoral degree. My husband was wrapping up things at Washington University. I had applied to and been accepted at several top chemistry departments. I had decided I would likely begin my studies at the University of Illinois-Champaign Urbana. There was only one problem….what about Mohammad…what if he could not get a postdoc or a job in Champaign? In the end, I decided not to go. But this was also influenced by another unfortunate event…the death of my father.

So my return to graduate school would ultimately be delayed three additional years as we sold my family home of almost three decades. Mohammad and I subsequently married and after 2 years of marriage our son John was born. I restarted my graduate studies at the University of Missouri-St. Louis when our son was 9 months old. I was 29.

Initially, I was very apprehensive about returning to school. I wasn't sure if I could keep up with my peers. I was concerned that I might have forgotten many things that I had learned both in my bachelor's and master's programs. I was my doctoral advisor's second graduate student. He was a new, untenured Assistant professor with a lot of enthusiasm for organic chemistry.

We hired a retired lady to come to our house and stay with John while I was at school and my husband was at work. Although it took most of my teaching assistant salary from the university, my son was well cared for and we had greater flexibility than if we had to take him to day care. I had to get my work done within a normal 40 h week. I learned to utilize my time at school effectively. Most of my studying was done at night after everyone else went to sleep as I would spend the evenings with my family. Cumulative exams were held on Saturday mornings once a month. On the Fridays before I would pull an all-nighter studying for my exams. Things worked out just fine, in spite of losing a little sleep. I excelled in the program, achieving straight A's in my course work and being the first student of my entering class to pass my cumes.

If I had to go to school on the weekend, I would try to save that time for running NMRs so my son could come with me and draw his own NMRs with the colored pens in the NMR room. Otherwise, my husband would watch John. Unfortunately, we did not have family members who could help us out with babysitting.

At my graduate school, I was surprised that there was only one female faculty member in my doctoral department (out of 21 faculty). After all, there seemed like there were a fair number of female graduate and undergraduate students. In my entering graduate class at least a third of us were women. The other graduate students said this female faculty member was very tough and that she didn't like it when female graduate students were too passive. Eventually I had this professor for a course and got to know her by going to her office during office hours for questions about the course material. She wasn't as scary as people had made her out to be; in fact, I quite enjoyed talking to her. When I think back on things, I think she was just trying to prepare us for what lay ahead. I continued to interact with this female professor throughout my time in graduate school. She learned about my interest in a future career in academe and would often put articles in my mailbox about the glass ceiling for women in academic chemistry. I was a bit hurt at the time, when she told me I might do better in industry than in academe. Now I understand what she meant; as I am now, even in this day and time, the only female in my department.

In the meantime, my husband had done two postdocs in the area. He was offered a "real job" at a small start-up pharmaceutical company in NJ during my fourth year of graduate school. He moved to NJ and my son and I stayed in St. Louis. In my fifth year, when I was writing my dissertation, my son moved to NJ to live with my husband. It was difficult being apart from both of "my boys" for that year, but it forced me to focus on finishing my graduate studies.

The NYC area proved to offer a variety of postdoc positions for which I applied. I defended my dissertation and moved to New Jersey over Labor Day weekend. I started working the Tuesday after Labor Day at Hunter College/CUNY. My postdoc involved a commute into Manhattan every day.

I would catch the bus at 7 AM every day and often would not get home until 9 or 10 PM. My husband's work on the other hand was only about 15 min away from home. My son attended kindergarten and elementary school near home so my husband would pick him up every day from after school care. I would do my best to make it home for school concerts, back to school nights, etc. My husband really held down most of the childcare responsibilities during this time; his help and support were critical.

I started looking for academic positions after about 2 years in my postdoc. Again, I was geographically limited as I did not want to live apart from my family. I would apply for any openings I saw in New York and New Jersey. Sometimes I would even apply for positions in Connecticut and Pennsylvania, knowing that I would have to commute. I applied to all types of academic jobs, including those at predominately undergraduate institutions, though I knew I would not be happy at these types of schools. I had research in my blood and having only part-time access to a fume hood in the physical chemistry teaching lab was not going to give me much opportunity for that. I wound up staying in my postdoc much longer than I had planned—almost 5 years. Now the question was ... was I damaged goods because of my age ... for starting as an Assistant Professor in a doctoral program?

In December of 1998, I interviewed at Seton Hall University in South Orange, NJ. It is a private, Catholic university with both undergraduate and graduate programs (MS and Ph.D.). The Catholic environment was very reminiscent of my undergraduate and MS days at Saint Louis University. But I especially liked the fact that I would be able to mentor doctoral students. What I wasn't really cognizant of at the time were the difficulties associated with doing graduate research in a small Ph.D.-granting department, such as lack of resources and collaborators and difficulty getting grant funding.

Also, the start-up money for setting up my lab was not great. However, Seton Hall is also one of the schools designated to receive money from the Clare Boothe Luce (CBL) fund for promoting women in the sciences. As such, I was hired as a Clare Boothe Luce Assistant Professor of Chemistry. CBL paid my salary for the first 5 years I was at the university and also paid an additional 20% of my salary as a stipend that could be used in any way I deemed necessary (including for childcare). I used the money to help pay for lab supplies when I was first starting out. I am forever grateful to the Luce fund for this support; it really made a difference!

My new department had 13 other faculty members (12 men and 1 woman). The other female faculty member had a lot of physical problems and 2 years after I was hired, she died. I was unfortunately unable to go to her for help and advice. Some of the senior male faculty in my department tried to help me along, though this was very informal and sometimes, quite frankly, wasn't even obvious. There was a Women's Faculty Association at the university and I reached out to them for a female mentor. The mentor assigned to me was in the math department which was exclusively an undergraduate department. So my assigned mentor and I had different sets of issues. I think we spoke briefly on only a couple of instances. I also became very active on different committees and on the Faculty Senate. This enabled me to meet women and other faculty in other departments. I attended local Women Chemist Committee (WCC) meetings and met more female chemists at other universities and at local companies. At one of these meetings I met Valerie Kuck, a retired industrial chemist and an active volunteer for the American Chemical Society (ACS). Valerie had an interest in doing research to elucidate the reasons for the underrepresentation of women on the faculty at top-ranked chemistry doctoral programs. This was a problem I had been keenly aware of since I was a doctoral student, and was also interested in. We recruited two psychologists from Seton Hall, Susan Nolan and Janine Buckner, and began a collaboration that lasted almost a decade. To this day I ascribe this collaboration as one of the major factors that helped me survive the isolation I felt in my department. Though the research we were doing was not chemistry, and probably wasn't respected by some of my departmental colleagues for this reason, we were able to get several grant applications funded and published and presented papers in multiple social science and chemical education venues. In fact, I remember an off-colored remark at a faculty meeting about how "maybe we all should apply for grants for basket weaving." In spite of this disdain for our work, I believe this collaboration helped strengthen my tenure application package and definitely gave me a support network.

I received tenure in 2005; my son turned 17 that year. I went for my first sabbatical, the year my son left home for college. Looking back, I always thought that I would have more than one child, but I just couldn't find a way to manage it. About a month after I was offered the job at Seton Hall, I found out I was pregnant. I worried and lamented on how to handle the discussion with my new department, but I never had to as I miscarried a month later. I am in awe of those people who have managed to have several children and to have productive and successful careers. There are different formulas for working out a successful professor–mother balance and having other supportive people to help you out is key. You have to follow your instincts and not be afraid to do for yourself what you feel is necessary.

Since coming to Seton Hall, I have mentored more than 50 high school, undergraduate, and graduate students as well as postdocs and visiting scholars. I have graduated six Ph.D. students. I was promoted to Full Professor in 2012. I look back on the days when I was struggling to balance work and being a parent. How much I worried at the time about not being there for all of my son's school events and the guilt I felt. I am blessed to have a wonderful, intelligent young man for a son with a good head on his shoulders. After all, everything did work out.

Lessons Learned

You can't do it all. Marriage and child rearing require a partnership. When both spouses work, there has to be a mechanism to share household and child rearing responsibilities. In my case, my husband is not a good cleaner and only in recent years has he begun helping with the cooking. He was extremely helpful to me in terms of taking care of my son so that I could stay late at work or get school work done at home. For about the past 10 years or so, we have also hired outside help for cleaning the house and doing the yard work. The helpers only come every other week, but it makes such a huge difference. It is so nice to come home to a clean house and to have time to relax on the weekends and spend time with the family.

Pick your department wisely. Some departments do not have a children-friendly mentality. For example, at my university, all of our graduate courses and departmental seminars are held in the evening. This is done to accommodate our large number of part-time students. This makes it difficult when you have children and need to be at home in the evenings. Also, ask yourself whether many of the faculty (male and female) in your department have children. This may affect their views on your needs to be available for your family.

Don't limit where you look for mentors and support. If there is no formal network of support in your department or university, don't hesitate to find your mentors elsewhere. Get involved with university/college committees. Go to the faculty lunch room and meet people in other departments. Go to local and national ACS events and network.

Don't be afraid to say no. As a professor and a mother, you have plenty to do without being talked into all kinds of university, departmental, or community service. Pre-tenure, it may be difficult to refuse, but don't be afraid to step away from these jobs once you have done your service. Also, I think it is important that everyone does their fair share.

Don't beat yourself up. There will always be guilty feelings about why you were not there for this or that with your children. Simultaneously we feel guilty for not getting work done: not getting the paper submitted when planned, not putting enough time on that grant application, etc. Just do the best you can and realize there will have to be give and take.

Don't be afraid to do what is right for you. Listen to your inner self and do what you need to be happy. Don't worry about what others think . . .and remember *there is more than one possible formula to success and also more than one definition for it.*

John and Cecilia Marzabadi—Summer 2013

Main Steps in Cecilia's Career

Education and Professional Career

1982	A.B. Chemistry, St. Louis University, MO
1984	M.S. Chemistry, St. Louis University, MO
1984–1986	Research Chemist, Monsanto Agricultural Products Company, MO
1987–1989	Laboratory Technician, Washington University, MO
1989–1994	Teaching & Research Assistant, University of Missouri-St. Louis, MO
1994	Ph.D. Chemistry, University of Missouri-St. Louis, MO
1994–1996	Postdoctoral Research Associate, Hunter College-CUNY, NY
1996–1999	Adjunct Professor, Hunter College-CUNY
1999–2004	Clare Boothe Luce Assistant Professor of Chemistry, Seton Hall University, NJ
2005–2012	Associate Professor, Seton Hall University, NJ
2006–2007	Visiting Associate Professor, Harvard University Medical School, Boston, MA
2012–present	Professor, Seton Hall University, NJ
2013–present	Visiting Researcher, Sloan Kettering Cancer Center, NY

Honors & Awards (selected)

1993	University of Missouri – St. Louis, Dissertation Fellowship
Sep 2003/Aug 2008	ACS Project SEED Service Awards
2004	Manchester's *Who's Who Among Executive and Professional Women*

Since 2001 Cecilia has been involved in Seton Hall University's Women's Studies Program, for which she has an Adjunct Appointment and was Acting Director (2005–2006).

Taking an Unconventional Route?

Janet R. Morrow

The path to my current position as professor of chemistry at a large public university could be considered to be a conventional one. I became interested in biological and chemical research as an undergraduate college student. I pursued that interest in graduate school and received my Ph.D. in chemistry. I was a postdoctoral fellow in two laboratories prior to landing my current job. Along the way I got married and had two children. When I embarked on my career 24 years ago, it was less common for women to major in chemistry and very few moved into an academic position. This was especially true for women with children. When I look at my current working life and my path to it, I feel very fortunate to have landed in such a challenging and interesting job. There is nothing more rewarding than having a family and there is nothing more enthralling than doing research.

Remarkably though, being a female faculty member in chemistry at a research institution in the year 2014 is still not very common. I have seen many talented women decide not to pursue an academic position even though I thought that they might have been happy and successful in taking this route. I write this brief account with the hope that it will guide junior scientists in their efforts to balance work and family. I hope my story will encourage women to pursue a career in academics if that is their desire.

J.R. Morrow (✉)
Department of Chemistry, University at Buffalo, State University of New York,
526 Natural Sciences Complex, Amherst, NY, USA
e-mail: jmorrow@buffalo.edu

Beginnings

I had a wonderful childhood in southern California. My father was an electrical engineer who worked in industry. My mother was a nurse, although after the birth of four kids in four years, she practiced nursing only upon request by friends. She still has an encyclopedic knowledge of medicine and physiology and obviously takes much enjoyment in it. My father was a traditional working father, but he encouraged both me and my three brothers to study science or engineering. Over dinner and on vacations, my dad would question us about how we thought the physical world worked. He would ask us how we thought electrons might travel in different media and discuss what he knew about atoms and molecules.

My three brothers and I body-surfed, skied, climbed, and hiked our way through childhood. We spent hours in the backyard swimming pool playing Marco-Polo and also playing soccer. My family traveled throughout the western USA for vacation, camping and backpacking along the way. My father noticed my interest in ocean life and encouraged me to study marine biology, a trendy topic at the time because it was forecast that humanity would need to cultivate the oceans for food.

College

I went to the University of California, Santa Barbara (UCSB), to study marine biology. The only university I considered attending was the University of California. The UC campuses are renowned for science, and the tuition was a bargain at that time. My parents had to plan for having four children in college at nearly the same time and this was clearly a factor in the decision. It turned out to be a good choice for me because I was able to take a range of courses to try out different science and engineering disciplines as well as to start undergraduate research projects early on in my studies.

The introductory science courses at UCSB were large and I wanted a more individualized experience. On the recommendation of a friend, I started research in the laboratory of Professor Barbara Prezelin in the biology department during my sophomore year at UCSB. I loved the research and I enjoyed being part of a larger team of people with a focus on research. I spent a summer and a couple of academic years culturing dinoflagellates and attempting to isolate a type of chromoprotein that was involved in light gathering for photosynthesis. On one memorable field trip, we rented a boat from the Scripps Institute of Oceanography and collected samples of dinoflagellates along the coast of Santa Barbara. But my protein isolations and purifications were difficult, and I was frustrated by my lack of background on how to improve separations. I enjoyed thinking about science in terms of molecules and liked the way that chemists could propose solutions to my questions on protein purification and chromatography. I transferred to the chemistry department in my junior year and started two new undergraduate research projects, one in ultrahigh vacuum studies of platinum surfaces with Professor Arthur Hubbard. Then I moved on to Professor Peter Ford's group to study the photochemistry of platinum organometallic complexes. After I graduated with my B.S. in chemistry and was waiting to attend graduate school in the fall, I worked on the preparation and characterization of micelles under the direction of Professor Henry Offen.

The experience of working in four different research laboratories in different areas of chemistry/biochemistry influenced my approach to science in later years. Chemical research is becoming increasingly interdisciplinary and this early experience encouraged me to dive into new collaborative research projects later in my career. I planned to go to graduate school because I so enjoyed these research experiences. I toyed with working for a year at a job in the food industry, but in the end decided to go right to graduate school on recommendation of my advisors. Spending a year in industry might have been beneficial, but clearly given the relatively long period required for graduate school, it was best to get through with little delay given that I was sure I wanted to pursue a graduate degree. At that point, I had a steady boyfriend but was not ready to settle down. I was eager to enjoy my time in graduate school.

Graduate School and Postdoctoral Studies

I chose to attend graduate school at the University of North Carolina, Chapel Hill, based on the strength of the department in analytical and inorganic chemistry. At the time of sending out applications, I couldn't decide which of these topics to pursue. Once I got there, I decided to study organometallic chemistry in the research group of Professor Joe Templeton. I liked the combination of synthesizing molecules, coupled with spectroscopic and structural characterization of the molecules and molecular orbital theory. The big change in scenery and culture from southern California was a maturing influence. For fun, I learned to whitewater kayak and to rock climb. Chapel Hill was rich in music, especially Blue Grass and Irish music. I remember that time fondly. Research went well and the bulk of my work was published prior to graduation. This was a tribute to my graduate advisor who submitted manuscripts in a timely way. I wasn't savvy enough at the time to appreciate it, but graduate student publication rate and quality is one very important metric to look for in a Ph.D. advisor.

Before I knew it, it was time to move on. I started searching for a postdoctoral position to continue my love of research, travel, and the outdoors. I chose to apply for international postdoctoral experiences and was awarded a fellowship from the National Science Foundation to study at the University of Bordeaux for 15 months with Professor Didier Astruc, an organometallic iron chemist. I defended my thesis and got married in the same week. A few weeks later, we left for France to start my postdoctoral position. Prior to that we would honeymoon in the French Pyrénées and climb with members of the French Alpine Club.

Working in a French laboratory took some adaption on my part. I had to change my research plan according to the available resources which were, at the time, different than in laboratories in the USA. For example, the NMR spectrometers were limited to hands on for staff only, but there was an EPR spectrometer and instrumentation for electrochemical analysis of iron complexes that I could use freely. I also struggled to adapt to a different culture and language. All in all, it was wonderful adventure. My husband and I traveled widely throughout continental Europe during the long summer, Christmas, and spring breaks. During one climbing escapade in the Wilder Kaiser mountain range that took place after I had attended a conference on organometallic chemistry in Vienna, I fell 150 feet down a cliff face and broke several ribs and punctured my lung. I was rescued by helicopter and spent a couple of weeks in the hospital followed by a painful train trip back to Bordeaux. That took a few weeks' time out of my research projects, and convinced me to move on to tamer sports.

This period in Europe was full of formative experiences that would build my confidence and courage to tackle challenging projects and to work with people from different backgrounds and nationalities. One of the most challenging was a presentation at the end of my stay that I gave to a committee made up of French and German scientists. This was more stressful for me than my Ph.D. thesis defense because it had to be given in French.

After 15 months of adventure and hard work, I was ready to return to the USA. I wrote to several faculty members and had three positions to choose from. I chose to work with Professor Bill Troger at the University of California, San Diego (UCSD), based on research interests, and his reputation as a good mentor. I also considered the university setting. UCSD is a large university that I thought would have good professional development and job placement opportunities. Bill was a wonderful mentor and allowed me to sit in on courses and to initiate a new project in the area of bioinorganic chemistry. After 11 short months, I was on the job market. I did apply for a couple of industrial jobs, but then put most of my efforts into searching for academic positions. I wanted a position in a large university with an emphasis on research, similar to what I had experienced.

Janet in her climbing days at Chapel Hill

Life as a Faculty Member

I accepted a position at the University at Buffalo, State University of New York (UB). UB is a large public university with a comprehensive set of colleges/schools including a college of engineering and a medical school. I moved there with my husband who soon started his own small business in violin making, sales, and repair. One of things I did not appreciate at the time was that western New York was also an excellent choice for raising a family. The commuting time was short, the public schools were excellent, and the family-friendly atmosphere made it relatively easy to bring up a child.

Nothing quite prepared me for the challenge of my first faculty position. It always is a busy lifestyle, but the learning curve at the start is steep. I juggled teaching new courses, leading students in research, being in the lab myself, writing manuscripts and proposals, and giving presentations as a faculty member. I made the error of starting too many projects and spreading myself too thin. I naively expected that my students would be as motivated as I was. Management skills were not something I learned very much about in graduate school and had to develop as I went along.

The birth of my first child, Erin, at the start of my second year as a faculty member forced me to narrow my choices of research projects. I chose the most promising ones and focused on them more effectively than I had before her birth. At the time, there was no maternity leave policy and I took off 2 weeks from teaching after she was born. As the only young female faculty member in the department, I received some negative comments on the pregnancy. One of my colleagues told me flat out that "some things are more important than tenure." A graduate student asked me: "Are you going to quit?" Other faculty members were more supportive and even offered to babysit when I had meetings. I was given the semester off the following spring. This turned out to be of key importance for my research. I could spend more time in the lab and I traveled to conferences as well. This time for travel enabled me to develop key collaborations, one in academics with Professor Bill Horrocks of the Pennsylvania State University and one in the biotechnology industry with Dr. Brenda Baker of Isis Pharmaceuticals. These collaborations enhanced the scope of my research and garnered the attention that was needed for a successful tenure case.

Post-tenure

Our son Garrett was born a year after I got tenure. This time, I took off 6 weeks from teaching. Like his sister, he an easygoing and happy child and this made it feasible to travel with him and to bring him to work as needed. However with two small children at home, I traveled less during this period. Despite the reduced stress from having attained tenure, I remember this period as being especially tough to balance work and home life.

Of course one of the ways to influence departmental atmosphere is through administration. I was associate chair for six years. In this position, I participated in the administration of the department and interacted with university colleagues in other departments. During this period, I observed that the attitude toward junior faculty had improved. We made the effort to hire more women and now have five female faculty members out of 30 total. I have resisted taking on more administrative duties, because research is still one of the most rewarding activities for me. Interdisciplinary and collaborative research requires a large time commitment in order to stay abreast of new fields and evolving research projects. Currently my research has a heavy component in magnetic resonance imaging contrast agents that contain iron and cobalt and in the design of metal ion complexes that interact with unusual nucleic acid structures.

Reflections

If there is anything I would emphasize, it is to be determined and committed to your science, but be flexible and keep a sense of humor about getting through your day. Maintaining a balance between family and career is challenging. In my situation, neither set of grandparents lived close by, nor could we afford a nanny. So I combined mundane activities such as grocery shopping with treats when the kids were small. We would get helium-filled balloons and cookies at the store while we shopped at Wegmans. I would make up exams for them to work on while I was grading mine. I took them on business trips and combined these trips with vacation activities when I could. Erin, my pre-tenure child, traveled with me to conferences, university lectures, study sections, and international trips right through her teens. We joke that this travel and continually new and changing experiences contributed to her restlessness as an adult. At 24 years of age and just graduated with a master's degree in electrical engineering from Columbia University, she has already traveled to more countries than I have.

Both of my children did experiments with me in the laboratory on weekends. We put together chemistry demonstrations for their classes. During one of them, an unexpectedly large chunk of flaming sodium metal shot up right into the middle school classroom ceiling. The ceiling was fortunately fireproof. The neighborhood kids remembered me for that one. Several years later, my daughter published a research paper with me while she was an undergraduate engineering student. She came to appreciate my career from a different perspective when she entered college herself.

Did my career influence my family time? Of course, but this is likely the case for any career woman. I generally work on the weekends, especially to catch up on reading. During certain periods of her life, my daughter was vocal about my lifestyle and commented that I was not a girly-girl. When I asked her to explain, she said that I wasn't an avid shopper, and didn't fixate on decorating the house. I didn't rush out to the furniture stores to get the best deals. I didn't fuss over my kid's clothes and appearance as her friends' moms did. In my defense, I remember doing many girly things such as taking long shopping trips in search of the perfect dress for middle school and for high school dances. In any case, your life will inevitably be different than most people's lives because you are a scientist. Your kids will let you know that.

Finally, it is of utmost importance to find a good balance between work, family, and other interests. There have been times in my life when things seemed especially challenging and it was important that I could fall back on friends and family or an activity outside of science for a brief respite. As your children grow, you will develop new interests and activities to share with them. I feel very lucky to have balanced raising children with an academic career. I hope that my route becomes a conventional one for women.

Janet and her family on vacation in the Cayman Islands

Main Steps in Janet's Career

Education and Professional Career

1980	B.S. Chemistry, University of California – Santa Barbara, CA
1985	Ph.D. Inorganic Chemistry, University of North Carolina at Chapel Hill, NC
1985–1986	Postdoctoral Fellow, University of Bordeaux, France
1986–1988	Postdoctoral Fellow, University of California – San Diego, CA
1988–1994	Assistant Professor, University at Buffalo, NY
1994–2003	Associate Professor, University at Buffalo, NY
1996–1997	Visiting Professor, University of Rochester, NY
2003–Present	Professor, University at Buffalo, NY

Honors & Awards (Selected)

1985–1986   National Science Foundation Postdoctoral Fellowship
1994–1996   Alfred P. Sloan Fellow
2007–2009   National Science Foundation Special Award for Creativity

Janet is the Director of a National Science Foundation Research Experience for Undergraduates site at the University at Buffalo.

From the Periodic Table to the Dinner Table

Danielle Tullman-Ercek

Photo: Peg Skorpinski Photography

How Did I Become a Professor?

Not until I was nearly finished with high school did people suspect I would someday become a chemical engineer or professor. I had a natural aptitude and love for math and chemistry, and since chemical engineering combines these two subjects, in hindsight it was an obvious choice. I also enjoyed my high school job of tutoring others in math and science, and especially loved that indescribable feeling when I helped a student achieve an "A-ha!" moment: the dawning of recognition on their faces and the almost tangible clarity they suddenly seem to emit was a moment of mutual excitement for me as a teacher as much as it could be for him or her as a student. A teaching career was therefore a clear option by then as well. Without the aid of hindsight, my career plan follows my interests from earlier in my childhood.

D. Tullman-Ercek (✉)
Department of Chemical and Biomolecular Engineering, University of California Berkeley, 116 Gilman Hall, Berkeley, CA 94720, USA
e-mail: dtercek@berkeley.edu

Growing Up

When I was about ten years old, my parents took me on a visit to a log home company from which they were considering buying a vacation/retirement property. I saw how the structure was made, including the many options for corners, insulation, seals, and roofing. Someday I would design and build houses; I decided: I would be an Architect! Even at ten, I was never one to do something halfway, so I pored over planbooks and tried to learn how to draft by hand.

My uncle, a builder, encouraged me. With his guidance I drew up countless floorplans and elevations over the next few years. Even after years of practice, my elevations resembled the drawings of a talented kindergartner. It was probably obvious to everyone that I was not cut out for architecture. Nonchalantly, my grandfather, who was a mechanical engineer, suggested I also consider an engineering career, since it would require less artistic ability. Just before I started high school I was logical enough to agree that engineering was a better fit for me.

I transformed all the passion I had for architecture into exploring other careers, especially civil and architectural engineering, and attended all the local workshops and outreach events I could for aspiring engineers. The next year I had my first detailed exposure to chemistry, and it became my favorite class. I began to tutor others in the subject and quickly expanded to tutor students in any math or science course. I nonetheless remained undeterred from my goal of a civil engineering career until I took physics and found force balances far less intuitive than any aspect of chemistry. In this way, by the time I applied for college, I was set on a degree in chemical engineering without knowing what, exactly, a chemical engineer did. It was enough for me to know that it involved math and chemistry, and designing, well...something.

The College Years

My father was in the U.S. Air Force and as a result of his career my family moved all around the country. Perhaps due to this upbringing, I was fiercely independent and eager to go out on my own. I also wanted to see a part of the country where I had not yet lived, so I applied primarily to universities in the Midwest. I chose to go to Illinois Institute of Technology, a small engineering college in Chicago. IIT offered me a full merit-based scholarship, as part of their new Camras Scholars program, and that made my decision of which university to attend an easy one. Once in college, I never wavered from my decision to major in chemical engineering. IIT offered an excellent introductory course to make students aware of the careers available in the profession, and I was delighted to learn that chemical engineers were needed to design processes (I knew that all engineers had to design something!) to make almost every consumer product imaginable, including pharmaceuticals. I still wanted to use my talents to do something positive for the world, so making lifesaving medicine seemed like the perfect fit. It wasn't quite as tangible as a completed building, but it was close.

My engineering courses were difficult, and filled with both theory and practical examples. I remember coming out of my first exam in my Materials and Energy Balances course thinking that I was going to change majors because I had done so poorly. As it turned out, I did do poorly on an absolute scale, but so did everyone else; we all failed to appreciate at first that engineering is not a topic to memorize if you wish to do well—a common mistake for entering engineering students. I stuck it out, of course, and as I began to understand the importance of treating each problem as a puzzle to be solved from basic principles, I performed much better. I even served as an undergraduate assistant for some calculus and chemical engineering courses over the next few years and served as a tutor in the campus resource center for three years. These experiences served to enhance my love of teaching, and I began to strongly consider a career in teaching. At that point, I thought a professor was only responsible for teaching courses, a vision that would soon be corrected.

During my junior year, I was encouraged by one of my professors to try an undergraduate research project in the area of mass transport and polymers. The project was theoretical in nature (no experiments or bench work) and I struggled with what was my first attempt at solving an open-ended problem that drew a little bit on everything I had learned so far. The project was unrelated to my favorite area of biotechnology, but I loved working on a problem that nobody had ever solved before. With this experience, I learned that research was about solving current real-world problems in creative ways and did not resemble my standard homework problems. I also spent a lot of time with both the professor who advised this research and his group of graduate and undergraduate students. I saw how much time my professor spent mentoring everyone and thinking about research and realized for the first time that THIS was what professors did in the many hours they had each week outside of the classroom. I also learned about myself: after spending so much time on theory but surrounded by experimentalists in the group, I knew I would prefer a project that required laboratory experiments. I also had yet to try working on a biotechnologically relevant problem and set that as a future goal.

As graduation loomed, I considered the idea of a Ph.D. because I was so interested in someday teaching at the university level. The expense of graduate school seemed far too high—there was a reason I had gone to the university that offered me a scholarship, after all—and so I resigned myself to at least starting out in industry to earn money for a higher degree. I rationalized that this was for the best, because I would be gaining valuable experience to pass on to my future students. Then I received a flyer in the mail advertising a Ph.D. program in chemical engineering. The flyer highlighted the fact that Ph.D. students would have their tuition paid for and would receive a stipend on top of that. I wish I had saved that flyer because it certainly marked the turning point on my career trajectory. After speaking with my professors about it, I learned that most graduate programs in engineering were set up in this way. They expressed surprise that I didn't know this already—how many others were unaware? I promised to spread the word. Among my close friends in chemical engineering at IIT, about half ended up pursuing a graduate degree.

Graduate School and Beyond

I applied and was accepted to a program at the University of Texas at Austin with Dr. George Georgiou as my first choice for dissertation advisor. Dr. Georgiou was already well known for his protein and antibody engineering achievements and fortunately agreed to advise me. At the time, I chose the lab because the projects were interesting and I knew Dr. Georgiou was highly respected for his work. To this day, I cannot believe I was so lucky given the criteria I used. It turns out that it was much more important to my career that Dr. Georgiou is an excellent mentor and genuinely cares about the lifelong success of each of his academic children.

My dissertation project required the engineering of a system for transporting proteins across the inner membrane of bacteria. This began my long obsession with studying the transport of materials across cellular membranes, a topic that eventually formed the basis of my own research group. (This is not so far from what I envisioned as a freshman undergraduate student: I design the microscopic factories—bacteria and other microbes—that turn sugar into medicines and other useful chemicals, and make it easier to get the products out!) As I neared graduation, Dr. Georgiou and I discussed potential postdoctoral research areas and advisors. It was the early 2000s, and he told me that I should think about getting involved in biofuels and/or synthetic biology. I stared at him blankly, having not heard either term yet. Both, of course, became quite popular within the next few years and I marvel still at Dr. Georgiou's ability to stay a little ahead of the crowd in this way. I followed his advice and ultimately joined the laboratory of an up-and-coming synthetic biologist: Dr. Chris Voigt—another rather lucky decision, in hindsight. At the time I worked with him, he was at the University of California San Francisco, and pre-tenure. I watched as he molded the laboratory research portfolio from a rather disparate set of projects into a focused research group renowned for its ability to engineer genetic circuits and program cells for applications varying from nitrogen fixation to spider silk production. Dr. Voigt also taught me a suite of valuable skills for success in the academic world, including grantsmanship, the importance of understanding politics, and effective communication.

By this point, I knew academia was the right place for me. I would engineer living microbial factories, and I would have even more impact by teaching thousands of others how to do it. Less than a year into my two-year postdoc in the Voigt lab, I began to apply for faculty positions. I was invited for several interviews and (admittedly, to my surprise) received multiple offers. A lot went into the decision, as is always the case when deciding on a job, but I found myself at University of California Berkeley in Fall 2009 and feel incredibly fortunate to have had my haphazard path lead me here.

From the Periodic Table to the Dinner Table

The Tullman-Ercek Group, June 2013, in front of the UC Berkeley campenile

How Did I Become a Mother?

I have always loved children, and the thought of not raising children of my own never crossed my mind. I did, however, worry about the logistics of raising a family—will I find the right partner? When is the right time to start a family? Should one of us stay home to care for the children while they are young? I also wanted to prove I was self-sufficient before settling down, and I wondered if this would be possible. Could I wait until my career was established before marrying and having children?

In college, I met the man I would marry within the first month of arriving on campus. If you had told me then that he was going to be the love of my life, though, I probably would have laughed—he wasn't my type, or so I thought. Jim appeared to be a typical southern California surfer boy who lived in the moment and didn't take much seriously. Though we had many mutual friends, we rarely hung out together or even spoke. During junior year, we had our first real conversation—about life, and not just the weather or how classes were lately—via an online messaging service and had an instant connection. It turned out we had a lot of shared views of the world, and his ability to live in the moment was good for me; he taught me the benefits of relaxing once in a while. We dated for the next couple years, but I was still clinging to the idea that I had to prove I could do everything on my own and so when I went to graduate school I made the decision of where to go entirely on my own. I then told him he could follow me to Austin, but I would not follow him back to California. Not surprisingly, he returned to southern California after hearing my position on the subject. Living apart, I figured out that while I was more than capable of going through life independently, it would be a much more enjoyable life with him in it. He came to a similar conclusion, and when he lost his civil engineering job in the downturn that followed 9/11, he took the opportunity to move to Texas. From then on, he would have an equal say in our decisions. We married two years later and adopted a rescued cocker spaniel. It was the start of my new family.

Jim found a job with a company based in Austin, and as my graduation neared, we had to make choices about both of our careers. His boss had children with advanced degrees—one of his daughters was even a scientist working in academia—and he understood the challenges we faced as a dual-career couple. He offered to let Jim work from home, wherever we ended up. This made the decision about where to do postdoc much less stressful because we now had only a standard one-body problem.

Jim was the sixth of eight children by his father, and had dreams of an equally large family of his own. I negotiated him down to three or four children, but the "when" was still a problem. Jim wanted to have children early, and I wanted to wait until our careers were stable. I also felt guilty about going on maternity leave; my career and that of my advisor depended on staying ahead of the others in the field and having a child would slow this process down, affecting everyone on the project. Soon after moving to the Bay Area, I figured out that I would be in my mid-30s before my career was anywhere near stable, and my biological clock kicked in. I also observed that an assistant professor had more stress and demands on her time than a postdoc, and that it only increased as a person moved up the ladder. It dawned on me, for the first time, that the adage "There is no best time to have a child" is absolutely correct. And if there isn't a best time with respect to career, then I reasoned that there was no point to waiting until I was older, which can certainly lead to more complications biologically. We made the decision to start our family while I was still a postdoc and were confident we would make it work. The saying "We plan while God laughs" describes what happened next. We tried for over a year, with no luck.

Meanwhile, I was choosing between multiple job offers—one at UC Berkeley and the others all from universities on the eastern side of the country. Each had advantages in terms of the program—Option A would be a supportive environment with terrific students; Option B would offer the opportunity to collaborate with incredible researchers, and so forth. Jim was not especially excited about staying in the Bay Area, nor about the high cost of living there, and wanted me to take the offer from the university nearest to his mother, who had relocated to the east coast. My family also was rooting for us to move to the east coast, as most of our extended family still lives on the eastern seaboard. However, I knew that the position at Berkeley offered me the highest chance of success because the university and its chemical engineering department are ranked among the best in the world. This meant that I would be able to work with the most talented researchers—professors and students—and I would have access to an extensive set of resources to help me get started. More intangibly, the community simply felt like the best fit for me and my research field. After a few weeks of debate, Jim agreed that Berkeley would be best for me and our future family, and I accepted my faculty position offer from UC Berkeley with his full support.

That same summer, I found out I was pregnant. We had given up on natural conception by this point, so we were surprised but thrilled. The timing was not what we would have planned, as the baby was due just a few months before my agreed upon start date at Berkeley, but we appreciated the miracle for what it was. Our son was born in the spring of 2009. This life change triggered some involved conversations between Jim and myself about how I was going to balance a demanding career and an infant. Our strategy was (1) to have his brother, who lived with us at the time, serve as a part-time nanny and occasionally make dinner and (2) to have Jim always carry out duties such as errands, cooking dinner on weeknights, and taking care of most household chores (bathrooms were still my responsibility) while cutting back on his hectic business travel schedule. It was hardly the 1950s model of gender roles, but this division of labor worked remarkably well, and it was a good thing—after a relatively easy pregnancy and delivery, we faced one new challenge after another with the baby. None of the challenges were out of the ordinary, but combined made for many sleepless nights and far too many day-interrupting doctor's appointments. Fortunately things calmed down as our son grew older, and we felt blessed to have such a wonderful little person in our lives.

As much as we loved being parents, we also quickly adjusted our plans for future children; Jim and I agreed that two children would make our family complete. We welcomed our second child, a girl, in the summer of 2012. This time around things went more smoothly, though her personality was the opposite of our son's in nearly every way, which meant we had another steep learning curve to overcome.

Despite the challenges, being a mother is the most rewarding job I can imagine. After a stressful day at work searching for funding and troubleshooting experimental protocols, I come home to children that make me forget everything with a smile or hug. Anytime I have a rough moment at work, all I do is think about my family and I instantly feel better.

The Ercek family at a birthday party in February 2014

How Do I Balance Career and Family?

I am often asked how I am able to handle my tenure-track position and my young children. To be honest, it would not be any easier if I were tenured, so the real question is how I balance any demanding career and still make time for family. In fact, many of my female colleagues, especially those who are pre-tenure like myself, also decided to start their families already. It seems to me that even 10 years ago, women in academia were much less likely to have children pre-tenure, and a shift in the academic culture is finally not only permitting but enabling this change. Thus I think I have it pretty easy compared to other working moms. I have a supportive spouse, for example, and a job that permits some flexibility in hours. The University of California has mechanisms for stopping the tenure clock for all new parents (mothers and fathers) and even participates in a plan that provides emergency backup childcare both at home and while traveling, providing more options should a perfect storm of problems arise. Nonetheless, I admit that my schedule is extremely hectic and far from that of a 9–5 worker. Thus I outline below the balancing act that I call "Monday:"

My days always start at 6:30 am, when my one-year-old daughter wakes up with the precision of an atomic clock. We haven't needed an alarm clock for months! She also wakes up my four-year-old son, so I (or my husband, if he has a free moment) feed them and get them ready for daycare/preschool. Once they are satisfied, I will turn on an episode of Mickey Mouse Clubhouse to entertain them while I get myself ready for work. We leave the house by 7:30, and they are in daycare/preschool by 8:00 am.

I take the train to work and use the commute to respond to e-mails that were neglected the previous day. Barring a transit delay, I am settled in behind my desk by 9:00 am, just in time to review the notes for my lecture one last time. The hour-long class starts at 10 am, and I hold my office hours immediately following lecture. During the lunch hour, I try to eat with other faculty as often as possible, but sometimes cannot avoid dealing with any crises that arose in the morning instead. At least once every 2 weeks I also make time to have lunch with other female faculty members on the tenure track; we share successes, failures, and strategies for balancing work and family, and these lunches never fail to put me in a good mood even on a bad day.

At 1 pm, I have the first of my scheduled meetings with my graduate students. Each meeting revolves around a particular research topic and involves two to four graduate students and the same number of undergraduate researchers. We discuss the previous week's progress, or lack thereof, and troubleshoot when necessary. We also devise a plan for the next week. These meetings are the highlight of my work week; it is the time when I get to interact with all of my academic "children" and I love talking about and brainstorming ideas for their projects. It is also gratifying to watch them grow into independent researchers.

By 4 pm, if I am on time (a rare event!), the last of these meetings ends. These meetings with graduate students are only on Mondays, but other days are equally filled with activities such as attending committee and faculty meetings, serving on graduate student qualifying exams, advising undergraduate students about careers and coursework, or serving as a peer reviewer on grant proposals or manuscripts.

Typically, the 4–6 pm block of time is used to take care of any issues that arose during the day—perhaps finding someone to fill a hole in a seminar speaker's day-long visit, or working with my lab safety officer to make sure the Standard Operating Procedure for a new chemical in our inventory is complete and accurate. Occasionally, I meet with a student for whom I will serve on the qualifying exam committee, or with my teaching assistant to discuss a homework problem, and if I am lucky I can use this time to get advice from my more senior colleagues. I also go to an on-campus yoga class from 5 to 6 pm at least once per week; it helps me deal with the stress in my life. I am noticeably grumpier and more tense if I skip yoga for an entire week! I leave by 6 pm so that I can be home for dinner by 7 pm, again using my commute to take care of e-mail correspondence.

My husband typically picks up the children from their respective daycares and makes dinner for us all. My son says grace and does a round of "Cheers!" and we eat together. It is my favorite part of the day. My son might tell me about how he found five acorns on the playground that he saved for the squirrels, or how he didn't get any time-outs that day. My daughter will say "Mmmm!" with each bite, even though only about 10% of it ends up in her mouth. She will then dance in her highchair as my son hums the theme of Jeopardy, which is on in the background, and by then it is time to clean up. I do the dishes as my husband coordinates bath time. By 8:30 we are finishing up the nightly routine of eating dessert, brushing teeth, reading a story, and tucking into bed. By 9 pm, the children (and often my husband, as well) are asleep, and I will use my last hours of the day to catch up on my writing. At any given time I have two or three grant proposals in progress and four or five manuscripts to edit for submission to a journal for publication, and these hours are my most productive for working on these tasks.

I have had trouble falling asleep for as long as I remember, and reading works better than any sleeping aid. Each night I attempt to read a chapter of a "fun" book, such as a Jasper Fforde novel or James Rollins thriller, but invariably I fall asleep after just a page or two, always by midnight, and the routine begins again...

Side note: I write about my typical Monday, but I have found it is helpful to carve out an entire day to simply work on my writing tasks—writing grant proposals and manuscripts on my research—in order to accomplish these two vital tasks for my career. These writing days for me are usually on Tuesday and are held sacred on my calendar; if I did not treat them as such, the time would be gobbled up by additional meetings. Since the writing must still go on, that means it would have to happen on weekends, which in turn would mean much less time spent with my family—not a viable alternative. There are other ways to build writing into the workweek, such as daily writing hours, but I found that, for me, devoting an entire day to writing is the best way to ensure it actually happens.

I love my jobs—both motherhood and professorship. It may be obvious from the narrative of my typical day that I love these jobs so much that I don't leave time for anything else. Not counting Mickey Mouse and Jeopardy, I rarely watch non-recorded TV, and grocery shopping and social networking are squarely in the domain of my husband. But this is my (and our) preference; if I didn't want it this way I would have switched careers, or decided not to have a family. This is my life, and I love it. The funny part about it all is that until I was in each role, I didn't have any idea what I was signing up for. So in a way, I was extremely lucky—twice.

The Ercek family on a camping trip in fall of 2013 at nearby Pinnacles National Park. Hiking with the family is a wonderful way to truly turn "off" and reduce the stress of a hectic life!

Interview with the Author

1. How has deciding to start a family or having a family influenced your career? How has your career influenced your family?

My career initially strongly influenced our decision about when to start a family. Though we married relatively early in my time in graduate school, I felt strongly that having children during grad school would send a message that my career was unimportant. Few women in my program had children during grad school, and other students joked that it was the only way to be guaranteed you could graduate on time (because the faculty advisor would not want to pay a student that was out on maternity leave, presumably). Thus while in graduate school, I did not think I should have children until I had tenure. I revised this decision when I realized that a tenured professor has even less time and many commitments that cannot be pushed back due to maternity leave (obtaining funding for graduate students, or publishing papers in a highly competitive field, for instance). Ultimately, I had my first child just prior to starting my faculty position. Surely those around us felt this was not ideal timing, but we made it work and I would not do anything differently if given the chance.

My family continues to have an enormous impact on my career, as well. I am learning to be a mother at the same time I am learning to manage a group of young researchers, and despite the age difference and skills to be learned, there are many similarities in the two functions. More than that, though, my family requires time, which requires an efficient schedule and that I turn work "off" completely for a minimum number of hours per day. I cannot stay late one day to finish writing a proposal, or I simply do not see my children that day because I get home after they are asleep. I also rarely work at home in the evenings until after the children are asleep, both lessening my guilt at working and making the time much more productive and free of interruptions. Overall, having a family enforced set of times during which I could not be deeply thinking about my research, and (as taking a step back often does) this did wonders for putting certain problems in perspective.

2. **Did you have role models? Which examples were set for you in your childhood or while you were growing up?**

    I write of the influence my grandfather had on me in the narrative, but my grandmother was also an inspiring role model for me. She often recounted the stories of her time in nursing school, and of how she met my grandfather and planned their wedding while simultaneously completing her nursing degree. She loved especially to tell me about one of her professors, who chided her: "Are you working on your M-R-S or your R-N?" She maintained that she could do both, and did. She and my grandfather were together nearly 60 years and raised three children, and she worked as a nurse (often on the night shift) even when the children were young. They did not need the money, given that my grandfather was a mechanical engineer with a productive career designing sewing machine parts for Singer. She worked all those years out of a love for her job, and I plan to do the same.

3. **Did you take any leave to raise your kids? How did you negotiate this? Do you have any advice regarding the organization/negotiation of leave?**

    The University of California Berkeley has put in place a number of helpful policies to accommodate faculty. For example, women who give notice of maternity leave (of the six- to eight-week variety) are also given one year of teaching relief and, if pre-tenure, one year of tenure clock stoppage. I did not need to negotiate for these benefits and felt they were supportive enough that I did not require additional leave. For my second child, I also had the good fortune of giving birth in June, affording me a lighter load not only during the subsequent fall and spring semesters, but also in the summer months such that I did not officially return to full duties until my daughter was 14 months old.

The Ercek family in Istanbul, September 2012. Prof. Tullman-Ercek was invited to speak at a conference, and decided to make that trip double as a family vacation

4. **Have you come up against any significant obstacles during your career and how did you overcome these?**

   I have not faced any extraordinary obstacles thus far, though it is still early in my career. I have faced a number of challenges that I believe are typical of a person on the academic path. For instance, I was "scooped" multiple times, both as a student and then as an advisor, and had to figure out how to regroup and make the best of these situations. I also had to learn to be a manager, which is not an easy task for an engineer with no training in that area. For each of these, I found that discussing the problem with my colleagues is useful in coming up with a plan. They have much more experience and offer many ideas on how to approach such relatively common challenges. Speaking with them is also reassuring—I am not the first to face these problems, and that means they can be overcome!

5. **Is there anything you would have done differently or would not do again?**

   Everything I did to this point made me into the person I am now, and so I would not change this for myself. However, I advise others to have children earlier in their career, and in their lives in general, if it makes sense to do so. I learned that there is no perfect time to have children and so there is no point in waiting for such a magical moment to appear. I also learned that children do not necessarily come at the time you specify! Moreover, it is quite difficult to be a parent, career or not, and some negative consequences of aging begin much earlier than I ever suspected, so having children while one still has boundless energy makes that aspect of life a tiny bit easier.

6. What advice would you give to young women hoping to pursue a career in academia? E.g., while studying, when planning a family

The best advice I ever received was from my mother. As I child, I hated to get anything other than the top score on every test and constantly compared myself to my classmates. My mother probably correctly worried that myself worth would plummet when (not if) I was no longer the top student. In direct contrast to the trending parenting strategies of the 1980s, which seemed to be devoted to making every child aware that he/she is special, my mother went out of her way to remind me each time I brought a perfect test or straight-A report card home that "there are always going to be many other people in the world who are smarter than you, and that is okay." She was proud of me as long as I did my best and did not expect me to be THE best. This didn't change my instinctive overachieving nature and perhaps even drove me to work harder to prove her wrong. But it also had its intended effect: it allowed me to more easily accept those times when I fell short of my goal, and it helped me to live with the fact that while I am a perfectionist, I am not perfect. This, I believe, was crucial to survival as a graduate student because failure in research is a necessary step on the path to success. Unlike many brilliant scholars, I knew how to accept failure and keep trying before ever setting foot in the research lab.

The mantra is easily reapplied to other aspects of life, such as motherhood. For instance, I know there will always be better mothers than me, so I cannot worry about it if I don't create a theme for each birthday party or if I cheat and use premade cookie dough to make Christmas cookies. As long as my children know I love them, that is all that really matters.

Message from Danielle's Son

My son is only four, but told me this: "My mommy is a very good teacher. I like to go to her office because it is fun to see her students. [He comes to the campus a few times a year, and whenever we do outreach events for families.] I also like to draw on her chalkboard."

The Tullman-Ercek lab group at the Monterey Bay Aquarium in August 2013. They took time for their camping and hiking retreat to look at the jellyfish and sea otters, and Prof. Tullman-Ercek brought her son along as an honorary group member

Main Steps in Danielle's Career

Education and Professional Career

2000	B.S. Chemical Engineering, Illinois Institute of Technology, IL
2006	Ph.D. Chemical Engineering, University of Texas at Austin, TX
2007–2008	Postdoctoral Researcher, University of California – San Francisco, CA
2008–2009	Postdoctoral Researcher, Lawrence Berkeley National Laboratory, CA
2009–present	Assistant Professor, University of California—Berkeley, CA

Encounters of the Positive Kind

Michelle M. Ward

> In everyone's life, at some time, our inner fire goes out. It is then burst into flame by an encounter with another human being. We should all be thankful for those people who rekindle the inner spirit. ~Albert Schweitzer

The above quote by Albert Schweitzer was the closest succinct way to encapsulate what I am hoping is the message of this chapter. I was more than humbled when I was approached to contribute something to this fabulous idea for a book, as I feel all that has been accomplished in my life can be attributed to traits I have learned from one or more persons I have had the privilege of meeting and working with. There are too many people who have impacted my life in some way to ever do justice to recognizing them all. What I hope to accomplish here is to layout some of the key lessons learned along the way.

Of course, my greatest encounter occurred in 1994 when I met my son, Zachery. He has given a level of beauty to my life that could not be matched by any other experience. I am proud of the different roles I have held over the years, and look forward to additional ones ahead, but for me the greatest title I hold is "Mom."

M.M. Ward (✉)
Department of Chemistry, University of Pittsburgh, Pittsburgh, PA 15260, USA
e-mail: muscat@pitt.edu

Just Do It

It may seem a bit cliché, but my parents taught the first key lesson learned to me. I did not come from a family where college was an expected path to be taken, let alone complete a Ph.D. My father was a machinist in a steel mill and my mother was a waitress. The lessons they imparted through their example, though, were the most valuable throughout the years. From my parents I learned the value of hard work and perseverance. I think most people in the sciences know this, at least the successful ones; however, I need to give it the credit it is due, as without the work ethic my parents provided me with, I would not be where I am today.

When I decided I wanted to attend college, I found a way. I didn't really have a mentor to help me with the process, but I knew that it was something I wanted to do and I found a way to do it. (The mentoring will come up later.) I worked two jobs to pay for my undergraduate schooling, on top of the student loans that were taken out. With the students I advise these days, and how busy they are even without outside jobs, I honestly have no idea how I pulled that off back then—aside from the innate programming to just keep moving forward and do what needed done.

I was married young and had my son almost two years after I was married. I was in the last semester of the senior year of my first undergraduate degree when he was born. I lived too far away from home to have help from family at that point, and he was too young for daycare, but I decided I would still finish that semester. Two weeks after he was born, I returned to school full-time (along with he in a stroller—I was very fortunate he was a quiet newborn) and completed my courses. Again, I just kept moving forward.

I had been teaching high school chemistry for four years when I came to the conclusion that I wanted to pursue a graduate degree. Unfortunately, likely due to the early age of the marriage, my son's father and I had been divorced for a couple of years at that point. I ended up moving closer to my parents, and I started graduate school as a single mother the year my son started kindergarten. There were multiple hurdles and key helpers throughout the experience (some of which I will touch on shortly), but I kept moving forward. One of my proudest days was when my son showed up to my dissertation defense and sat in the front row. He didn't really understand what I was talking about, but he said he was proud.

I have had the opportunity to mentor and advise several young mothers in my department. I, of course, have lots to say to them, as I would love for them to have a more direct path to their goals. The one thing I keep going back to is to just keep moving forward. If there is something you want and you keep working at it, even if it is just by sheer will, you can get there. In my mind, perseverance and hard work cannot be emphasized enough as a key trait for any type of success.

Embrace Serendipity

I didn't really have any opportunities to interact with a female role model in science until I transferred to the University of North Dakota. That young marriage I mentioned earlier resulted in my transferring from a very small school to this larger school with a graduate program. At this new school, I encountered my first female chemistry professor, Dr. Kathryn Thomasson, who also introduced me to the world of research.

I am a huge proponent of all types of diversity in departments. I don't think we should compromise pursuing the most qualified candidates, but to perhaps rethink how one attracts those who are qualified AND also increase the diversity of a department. There are many books and studies out there that can provide outstanding data on this topic, and I am by no means a diversity expert. From my own experience, though, I can say having a female professor (someone who "looked like me") opened the world of academia being a possibility. I did end up pursuing teaching high school for four years after completing my undergraduate degrees, but I believe seeing a woman in academia allowed me to more readily see myself in graduate school when I did decide to go down that path.

Transferring to the larger school also resulted in my changing from an Education Major (with a focus on chemistry) to both a Chemistry Major and an Education Major. The requirement of the additional upper level chemistry courses, and the exposure to research, resulted in an easier transition to graduate school than I would have had otherwise. The interesting part of my undergraduate research had to do with the fact that I was to complete it while I was pregnant with my son. Kathryn facilitated this, and kept me on as a work-study student after the coursework was completed, in an extraordinary way. I was drawn to the theoretical work she did, due to my circumstances and my interest in physical chemistry at the time. A good portion of my research was conducted with my son in a playpen beside my computer desk, at her recommendation. The work I did with her resulted in a couple of publications, which assisted my application process considerably for graduate school. At the time, I honestly had no idea how many doors would be opened due to the good fortune of working with this woman and her open-mindedness regarding my situation.

Don't Be Afraid to Ask

I thought about titling this section "Science Dad," as a huge part of what I have been able to accomplish I owe to Dr. Sanford Asher (Sandy); however, there are some others who have provided very complementary lessons. Learning to not be afraid to ask for what I want or need has changed my career—and my life. Sandy, and others, taught me that I would not get what I want in a timely manner if I don't ask for it and realize I am deserving of it.

I met Sandy when I was applying to graduate program at the University of Pittsburgh. I was interested in his research, but was intimidated by the time requirements of the program and originally turned down my offer to attend Pitt. One evening, a few days after I sent in my refusal letter, I received a phone call at home from Sandy asking me to come back in to the department and discuss my decision. We discussed my career goals, and he explained a concept that was foreign to me ... to teach at the University level, it was not about teaching training as much as it was about the quality of your research. (I certainly think that some professors could improve from some actual educational training, but that is a topic for another book.) Of course, I didn't want to limit my future career opportunities, so it made sense to attend a program like Pitt had ... but there were issues regarding my situation with my son. As I mentioned earlier, I was going to start graduate school as a single mother when my son was about to start kindergarten. At the prodding of Sandy, I disclosed I was worried about things like parking on Pitt's campus (a nightmare for anyone who has experienced searching out a metered spot there) and the time I would need to spend working in the office when Zachery would be at home. As a result, Sandy arranged for the Department to provide me with a parking pass and a home computer. I could arrive on campus with no issues after dropping my son off at the bus stop, and I could access online resources to do necessary desk work from home. This was the first time I was shown to ask for what I wanted, instead of assuming I would have to compromise or find a work-around. You may not always get what you ask for, but you are certainly less likely to get it if you never ask.

I owe the current position I hold at the University of Pittsburgh to interactions with two main people, Dr. George Bandik and Dr. Joe Grabowski. As I was nearing the completion of my Ph.D., I was only pursuing positions in industry. I had fallen in love with research over the years, and as I was geographically constrained to the Pittsburgh area, I knew the type of tenure-track academic position I would want was not going to be a possibility. On multiple occasions, George would talk to me about considering applying for the Lecturer and Laboratory Coordinator position that was available in our Department. I would talk to him and others about not wanting to leave research to "just teach." George never gave up (thank goodness for stubborn people), and Joe really made things hit home when he pointed out the fact that the interviewing process would be about me interviewing the Department as much as them interviewing me. I realized there was no reason to see the position as a "lecturer box" that I would have to conform to and that I would never really know what was possible for this position until I pursued it. There are certainly compromises that have been made, but I have been able to make this position into something extremely fulfilling, and, to be honest, I feel the Department is better off for my not wanting to be "put in box" and asking for what I wanted.

"Peacock-ing"

It might sound simple, but I thought that if I worked hard at something it would be recognized without me having to point it out. I also thought promoting myself and my accomplishments would be viewed in some sort of arrogant manner. In recent years, I have started to read many articles and books geared toward the art of negotiation, specifically for women. It turned out I was far from alone in my thoughts that I would be recognized if I waited for my pat on the back, but more so it became very clear that recognition would not come in an expedited manner without learning to be my own promoter.

Once again, I have to return to the guidance of my "Science Dad." When I first started attending national conferences, or even somewhat audience limited grant progress report meetings, I was in awe of the scientific talent whose presence I was in. Sandy would make a point of inviting his graduate students to sit at the lunch table with himself and these prominent scientists. Eventually the conversation would turn to research projects going on in our group, and somehow (very often) the assumption would be made that the male graduate students sitting at the table would be the lead students on those projects. Sandy was great about very respectfully letting those scientists know that I was the one leading the project and then encourage me to discuss what the state of the research was. Through these types of experiences with him, I learned to not pause and wait for the conversation to turn to what I was doing and to not be hesitant in letting those at the table know of my accomplishments . . . there is a way of doing these things without being arrogant and to earn the respect of those at the table, so to speak.

When I was at the point interviewing for positions toward the end of my dissertation work, I was very fortunate with job offers. I had worked hard for the credentials I had and was ecstatic, to say the least, at the thoughts of being on the verge of not living on a graduate student stipend and being able to step into my own. Sandy was my first introduction to negotiation—and the idea that I was worth more than the initial offer. In addition to learning how to ask for what I wanted or needed to increase my opportunity for success while raising my son, I learned to be willing to ask for more to benefit my career based on what I had to offer. At the time it seemed a bit foreign to me to have to point out that I was deserving of and could ask for more than what was first offered, and at times I still have to remind myself of this. Not every job offer will provide you with everything you need, but unless you are willing to promote yourself and confidently go after what you want, every company/institution would be happy to take you at a bargain.

In my position with the University of Pittsburgh, I am not a tenure stream faculty member so I go up for contract renewal on a somewhat regular basis. The first time I was up for this renewal, Dr. Adrian Michael taught me the idea of having to "peacock"; this was not his term, but it succinctly grasps what he took the time to share with me. I was asked by the Department to list the activities and accomplishments I had completed over the year. Being a scientist, I made a list of bullet points, for the most part . . . clearly everyone reading the list would see the importance and magnitude of what I had tried to accomplish, in my mind. Under the guidance of Adrian, I have come to realize that is not always the case. Even if you already have your job and work with supportive colleagues, it is one's own responsibility to be sure you receive the credit that is due the hard work put in. It does not benefit your career to wait and hope that others see your worth ... it is your responsibility to point that out. Reading the books mentioned earlier, it seems this comes more naturally for men, so I was very fortunate to have promoters in my corner who helped with this until it became more natural for me.

It's OK to Not Do It All

This might be the hardest thing for one to come to terms with; I know it was for me. In this day and age where more doors are open for women than ever before, we still must realize you cannot do everything you want to do 100%. There are always compromises that must be made, and one must make the choices that they can live most comfortably with.

I spent eight years completing my graduate degrees. I had to watch some students enter the research group after me and then leave before me. It was a blow to my pride and made me question myself at times, but at some point I came to the conclusion it would not matter in the long run if it took me longer than my classmates. I was a homeroom mom, I never missed one of my son's school plays or spelling bees, I was home to pick him up off the bus almost every day, and we had family dinner every night. When I went through the graduation ceremony and my family took me out for a late lunch to celebrate, my son offered to give our name to the hostess ... he put the reservation under "Dr. Mom." It really was at that point that I was able to let go of the frustrations for "taking so long" to finish my degree. I had done it in a way that meant more to me—I was a Ph.D. and a mom. I am so glad I didn't give up those opportunities to be a part of his childhood memories to finish a year or so earlier. I may have missed out on some opportunities while in graduate school, but I can live with that knowing I was there for him during that time which I could never get back.

When I was pursuing a position toward the end of my graduate work, I chose to accept the non-tenure-track position I currently hold. I know that I was capable of pursuing something more prestigious, and, again, it took me some time to come to terms with that compromise. But again, I never missed one of my son's soccer games, band concerts, formal dance pictures, and the like. I was able to arrange my schedule so that I was never off traveling, missing out on events I would never get a second chance to see. I was able to keep him in Pittsburgh and near my family. Now that my father has passed away, I am so very thankful I made that decision, as he and Zach were very close. I hold absolutely no judgment against those who would have chosen a more career-beneficial path—I just think it needs to be said, no one should judge those who put their child's needs above their own career goals. There is nothing wrong with finding the position that allows you to be the parent you want to be, and there are ways to contribute to the scientific community outside of a research-intensive career path.

A final thought in this category ... ask for help when it is needed. Needing and asking for help don't make one any less of a success. I am by no means a supermom, and I readily admit it. I tried to be there for my son as much as possible, but there were times I needed outside support. My parents were always willing to step in and help with him when I couldn't pick him up from his bus for various reasons. My brother was an amazing role model for my son and would stay with him when I would be off at conference. I knew many wonderful mothers at the school he went to who would help get him to or from soccer practices when I couldn't swing both directions. There is no way anyone can do everything by themselves and hope to be successful at everything. Accepting my lack of superpowers has allowed me to take pride in what I could do and to truly appreciate those in my life all the more.

Pass It On

It might seem strange, but one of the biggest advantages in my career recently has been taking a strong role in mentoring and science outreach. When I realized a tenure-track position was not in my future, I decided I would further contribute to the scientific community by focusing the time that would have been spent on running a research group on mentoring students and being involved in science societies that emphasize science outreach and excellence. I serve as an advisor to newly declared chemistry majors and have informally advised a multitude of undergraduate and graduate students over the years. I have a personal goal of trying to help others find their way, perhaps a little more straightforwardly than I did. I certainly feel the need to "pay forward" what was done for me by those I was fortunate enough to encounter along the way.

In addition to wanting to help others, the involvement in what I would consider scholarly service has helped continue to push me forward in my own career. I started a local Women Chemists Committee in the Greater Pittsburgh Area three years ago, with the hopes of providing increased mentoring and networking for the generation that comes behind me. What I didn't realize at that time was that as I was focused on finding ways to help these women around me, I was being forced to focus on what steps would help one succeed ... encouraging others around me to see their worth, to think outside the box, to best prepare themselves to go after their dreams has kept this in forefront of my own mind. I love being able to work with the young women and men I have the opportunity to and find it very fulfilling, but it has also helped me to stay on top of the actions that will ultimately help me be more successful.

Final Thoughts

As far as my advice for those women considering a position in academia and successfully navigating the idea of a life–work balance ... do what is right for you. Define your own path and make your own rules. Accept that nothing is perfect. Never sell yourself short.

Main Steps in Michelle's Career

1994	B.S. Chemistry, University of North Dakota, ND
1996	B.S.ED. Secondary Education, University of North Dakota, ND
1996–2000	High School Chemistry Teacher, Butler, PA
2003	M.S. Chemistry, University of Pittsburgh, PA
2008	Ph.D. Analytical Chemistry, University of Pittsburgh, PA
2008–2009	Postdoctoral Research Associate, Department of Chemistry, University of Pittsburgh, PA
2009–present	Lecturer and Laboratory Coordinator, Department of Chemistry, University of Pittsburgh, PA

Honors & Awards (Selected)

- 2013 Students' Choice Award from University of Pittsburgh College of General Studies Student Government
- 2013 Outstanding Service Award from University of Pittsburgh Department of Chemistry American Chemical Society Student Affiliates
- 2011 J. Kevin Scanlon Award for Dedication in Enhancing Science Education

Michelle is an active member in many associations, including ACS and Iota Sigma Pi National Honor Society for Women in Chemistry. She is the founder of the Greater Pittsburgh Area Women Chemists Committee.

Encounters of the Positive Kind 137

Celebrating with my "Science Dad" (Dr. Sanford Asher) after successfully defending my Ph.D. (2008)

Celebrating Zachery's high school graduation (2013)

Family time with my father, son, mother, and brother (2011)

The Long and Winding Road

Gail Hartmann Webster

I can't remember a time during my precollege education when I didn't think that teaching would be a great profession for me. I was always the kid who enjoyed helping my classmates with their work, or the children I babysat with their homework. I also remember getting very frustrated if I didn't understand something as quickly as I thought I should be able to. I can see all of these traits at play as I've made my way from high school through college and graduate school and back to the classroom as a professor.

I was very fortunate to go to a great high school in southeastern Virginia. My teachers were excellent and my classmates were too. At one point, I think I counted that there were at least five of us in my graduating class who went on to get Ph.D.s. During high school I was continually challenged by my teachers and my classmates to push myself academically. I took two years of high school chemistry "back in the day" before Advanced Placement courses were all the rage. My high school teacher was a former industrial chemist, and I'm sure she let us do experiments in class that would never be allowed anymore. But, it was fun, it was challenging, and I enjoyed it. I liked chemistry so much that I decided to go to college to be a pharmacist.

I attended Virginia Commonwealth University and completed two years of a pre-pharmacy curriculum and entered pharmacy school on VCU's medical campus

G.H. Webster (✉)
Department of Chemistry, Guilford College, 5800 W Friendly Ave., Greensboro, NC 27410, USA
e-mail: gwebster@guilford.edu

in my third year of college. It wasn't what I expected. I sat in a room with about one hundred other students, and the professors changed every hour or so for the next class. Occasionally, we had breaks in the day, and there were some labs, but what I remember most is sitting in the same room for most of the day and then studying like crazy at night for tests. When the summer came, I did a required externship (essentially an internship, but not supervised by the college) at a local pharmacy near my house. It was an awful experience for me. I disliked the job. My particular experience didn't give me hope that it would be a profession with much room for intellectual growth. I left pharmacy school and went back to VCU, but this time, I was an education major and a chemistry minor. It was during this time of transition that I met my husband, Jeff. He was working as a researcher at the medical campus of VCU and he came to see me over that summer and helped me get through a very confusing time.

I took more chemistry classes than required for chemistry minor. I took the full year of physical chemistry that was not required and I also did a year of undergraduate research, which was a major influence on my career path. My last semester of college was supposed to be spent student teaching. I say "supposed" because I was asked to leave student teaching early and head to northern Virginia to interview for an unexpected opening for a chemistry teacher. A teacher left her position on disability and the students were experiencing a revolving door of substitutes. I was doing well in my student teaching, so my faculty advisor recommended me for the position and suggested that I interview. I was offered the position and VCU let me leave my classes early that semester. I became a full-time chemistry (and physics) teacher. I graduated from VCU in May and finished teaching my classes in June.

My year of undergraduate research with Sarah Rutan at VCU was instrumental (pun intended) in my decision to become an analytical chemist. I thoroughly enjoyed working with her and the graduate students in the lab. Sarah encouraged me to present my work at meetings, and more importantly, she encouraged me to apply to graduate school. In the meantime, Jeff and I became engaged and when grad school entered the picture, we decided to apply together (him for biochemistry and me for analytical chemistry), then see where we got in, get married, and move on with our lives. I think Jeff was even more insistent that I apply for graduate school than Sarah. A graduate degree was never in my original plans. Neither of my parents went to college, and just getting a B.S. degree was good enough for me. Looking back, I realize how lucky I was to have a professor take so much time and energy to help me realize I had the ability to get in and get through graduate school. I also know that I am even more fortunate that I found Jeff and his unwavering support.

Jeff and I entered graduate school at North Carolina State University two weeks after we were married. I'm not sure if I'd recommend that for anyone, but that's what we did! My lack of confidence resulted in my entering grad school in a master's program. Jeff knew he wanted a Ph.D., but I wasn't so sure. I found a great research group to join, but my advisor never had much grant money. All of the grad students in my group were teaching assistants, and for me, being a teaching assistant was the best part of graduate school. At NC State, I had some wonderful mentors who knew I wanted to teach and supported me on my path. I remember Dr. Forrest Hentz telling me one day, "Gail, you've got to get your union card if you want to teach, and it needs to have three letters." Well, I decided that IF I could get my M.S. THEN I would stay on and get the "union card." I did, but it took me longer than most folks because I spent a good amount of time teaching, tutoring, and running a supplemental instruction program in addition to doing my research.

Meanwhile, my husband stayed focused in the lab, finished his degree, and started a postdoc at the University of North Carolina in Chapel Hill. He had some great offers for postdocs at some incredibly prestigious places, but since I wasn't finished yet, he stayed in the area for me. I am a lucky woman! At this point, I was getting anxious to start a family. I became pregnant, had a miscarriage, and became awfully discouraged. Several months later, I got pregnant again, and this time, there were no issues. I finished my last few experiments, had a baby, and wrote my dissertation. Caroline Webster entered the world the February before I graduated in May. Sometimes, I don't know how I ever finished. But, then I remember the days of writing like a fiend when she was sleeping, which wasn't often those first few months...boy was she colicky! I also remember Jeff coming home from UNC, us having a quick dinner, and then him going back to NC State with me at night while I sat at the computer in the lab and worked. He would walk laps around the building with the baby while I typed and did data analysis.

I defended my dissertation when Caroline was three months old. I nursed her just before I went to do my seminar and have my final oral exam, and she promptly threw up all over me. One of the departmental administrative assistants noticed the barf all over my shoulder and down my back and alerted me of the situation. I told her, "Well, maybe they won't keep me quite as long!" What could I do, really?

Gail's Ph.D. graduation from North Carolina State University, Raleigh, NC in December 1994. Jeff Webster is the proud husband and father, Gail in the regalia, and Caroline Webster, 10 months old, is in her Wolfpack Red Christmas finery

Just after Caroline was born, I saw an ad in our local newspaper for a teacher at the North Carolina School of Science and Mathematics (NCSSM). NCSSM is a public, residential high school for academically talented juniors and seniors. With Jeff in the middle of a postdoc, with pharmaceutical companies in Research Triangle Park downsizing, I was pretty limited about what I could do next. None of the big universities in the area would hire me to do anything other than teach labs, and postdocs in analytical chemistry in the area were scarce. So, I applied for the job and started teaching in the fall.

It wasn't easy being a working mom, with a postdoc husband, and no family nearby. We bought a house a few years into graduate school, so with a mortgage and day care, money was pretty tight. I found a great in-home day care for Caroline near my work place. I commuted about forty-five minutes each way, so I had time in the car with her. My job at NCSSM was a great experience. My students were engaged and fun to be with. My colleagues taught me more about running a general chemistry lab program than I ever learned in graduate school or from my teacher education program in college. The difficult thing was what to do when either Caroline was sick, or the babysitter was sick. NCSSM didn't have substitute teachers, and we had no family nearby to help out. Jeff was always the one who stayed home with the sick baby, and he'd go in at night or stay later to make up for his lost time. There were several times I had to take Caroline with me to school, and I often got some of the students to help me with her. I don't think my colleagues approved, but I did what I needed to do to keep my job and to help Jeff make progress in his work as well. I always kept my eyes open for other opportunities too.

I taught at NCSSM for two years. Each summer, I applied for a summer program for high school teachers at the National Institutes of Environmental Health Sciences (NIEHS) a division of the National Institutes of Health located in Research Triangle Park, NC. It was an opportunity for me to continue to be research active when I didn't have many other opportunities to be in the lab. I worked at NIEHS for two summers doing molecular biology. In my second summer at NIEHS, my graduate research advisor called me and told me that NC State had a job that had "Gail Webster" all over it. The chemistry department was hiring two teaching faculty to work in the general chemistry program and work toward reform in the chemistry curriculum. I wanted this job. However, I knew I was pregnant. I thought I should just let this opportunity go, and be satisfied where I was, even though I wanted to work at the college level with all of my heart. I didn't know what to do and I remember talking to my mom about it. I clearly remember her saying, "Gail, it's 1996, women have babies and work. If you don't apply you'll never know if you can do it." So, I applied, I interviewed while I was twelve weeks pregnant, and I was offered the job. I might add that I was offered the job even though I was told that the department chair said he would never hire an NC State graduate. As a part of the interview for the position, I also had to do a research seminar for a 100% teaching job, so my work at NIEHS was essential for me to be able to give a presentation on something other than my graduate work.

Working at NC State was great. I was closer to home, Caroline was in day care near Jeff's work (his advisor moved from UNC to NIEHS), and I loved teaching chemistry at the college level. Those professors at NC State who had been great

The Long and Winding Road

mentors during graduate school were continuing to be mentors at the beginning of my career. When teaching schedules were being decided for the spring semester, I was in a bit of a quandary. I had to tell folks I was pregnant, since the baby was due at the beginning of the spring semester. When I told my direct supervisor, I will never forget what he did. He threw a pencil across his desk and said, "I knew this was going to happen." He told me I was responsible for finding someone to teach my classes. He was clearly angry and didn't give me any support. I was upset. Here I was, the first teaching faculty member to have a baby in the chemistry department at NCSU...in 1997. I asked to be given a night teaching assignment and told him I'd be back at work as soon as possible after the baby arrived. I was very proud of myself for keeping my cool during this conversation. As soon as our meeting was over, I went to my Ph.D. advisor's office, walked in, shut the door, told him my story, and I cried. He was so calm and said, "Gail, this is academia. We help each other. I'm happy to take your courses while you're gone." And so he did. I'll always be thankful to my advisor and mentor, Chuck Boss, for helping me when I needed it through graduate school and in my role as a visiting assistant professor.

Rebecca Webster was born in late January. I had two weeks at home with her, and then I started back to work teaching my evening class. I regret not having more time with her at home to this day. No woman should ever feel like they have to go back to work so quickly after giving birth. It was difficult, and Becky had jaundice soon after birth so I was really worried about her health. It was not an easy start for us. I was fortunate that instead of teaching two classes, my load was reduced to one class that semester. I would arrive at work with Becky in a stroller, and Jeff would meet me in the lecture hall on his way home with our two year old and take both girls home with him while I taught my class. I have to say, I had the most amazing group of students that semester. They sent me so many kind notes and e-mails, that I'm still touched by their kindness.

Gail and her husband with their daughters Rebecca (five months old) and Caroline (three years old). This photograph was taken in the summer of 1994 in Corolla, NC in the Outer Banks

Later that spring I interviewed for the coordinator of general chemistry labs at NC State, and I got the job. I moved into an administrative position and I wasn't teaching anymore. It wasn't my ideal job, but I needed something more than a one-year contract. Jeff was coming to the end of his postdoc and he was starting to interview for jobs. I realized that he had a greater earning potential than I did and I was willing to move wherever we needed to be to take care of our family. At the end of that fall semester we moved from North Carolina to Delaware for Jeff to begin his job at DuPont Pharmaceuticals.

We moved in December, after I made sure the teaching assistants turned in all the grades for the many sections of general chemistry lab. Jeff moved to Delaware before I did. I had to finish up the semester, get the house on the market, and get us moved. Not as easy as it sounds, but I did it. I had no job, but I was able to spend some great time with our girls. While I wasn't scouring the job listings, I was looking a bit. I found a temporary teaching assignment in Philadelphia, about thirty minutes from our home in Wilmington, DE. I worked there for a year, but with two children in day care, the expenses didn't justify the work. I left after a year and was at home again with my girls.

The October after I left my job in Philadelphia, Bristol Myers Pharmaceuticals bought DuPont Pharmaceuticals. Massive layoffs occurred, and my husband was out of work, and I wasn't working either. I felt awful that I did not have a full-time job to help us through such a difficult time. I found a part-time job tutoring, but I knew I needed to make sure I never put us in such a position again.

In April, Jeff started a job in High Point, North Carolina, at a small biotech company. I followed in June after Caroline finished second grade. Again, I got the house sold and got us moved back to North Carolina. I stayed at home the first year we were in North Carolina. Caroline was in third grade and Becky started kindergarten. I was able to volunteer in their school. I got in touch with some of my former colleagues when we moved back. One kept sending me job announcements and encouraged me to start applying. She sent me a job posting for teaching summer school at Guilford College, a small liberal arts college in Greensboro, very close to our home. I applied and got the job. I started teaching at Guilford College in the summer of 2003 and I've been there ever since, but it hasn't been easy.

I began working at Guilford as a visiting assistant professor in summer school. I let my colleagues in the department know I wanted more, and I was offered a job as a lab manager (part-time) and as a part-time instructor, teaching two classes the following fall. My two part-time jobs did not add to a full-time job, but I did it anyway. I did it because my colleagues in the department were awesome. They were supportive, they offered help, they listened when I had ideas, and teaching was truly important. My teaching job in Philadelphia was the opposite. I honestly never thought I'd teach again after that year. I was told so many times at my former job that teaching was important, but when it came time to hire, the only consideration was research, which was not my strength.

After a year as the part-time lab manager, the college hired a full-time lab manager and moved me to a full-time visiting assistant professor. I worked as a visiting assistant professor until 2006. That August, I began a tenure-track position. I changed my area of research from analytical chemistry to chemistry education, and I received tenure in 2010. From 2003 until 2010 Jeff was my anchor. He helped out with the girls more than anyone can imagine. He went to all of the programs at school. He took time from work when they were sick. I was never the one to stay at home with them or to take them to the doctor. He ran them around to more practices and early evening activities than most dads ever would. I missed Becky's kindergarten graduation, I missed both girls' fifth-grade promotion ceremonies, and I missed Caroline's eighth-grade graduation ceremony. I gave up a lot to prove that I was worthy of tenure. 2009 was a tough year for us. The biotech where Jeff worked started doing layoffs and he lost his job in January 2009. The recession was in full swing, and he didn't find a job until January 2010. The job was in Seattle, Washington.

Since I was waiting to hear about tenure, I didn't want to give up my job and move to Seattle. I stayed in North Carolina with the girls, now in tenth grade and seventh grade, and tried to keep things together while he started his new job at a contract research organization. It wasn't easy raising two adolescent girls, keeping up with their activities, working full-time, taking care of the house and doing all of the professional things one needs to do as an academician. Jeff got back to North Carolina once a month to see us, and the first July he was in Seattle, the girls and I flew out and visited him for a month. We never seriously considered moving across the country, although there were times I wanted to throw in the towel and move! It's a good thing we chose to tough it out. In 2013, the site he worked in was shut down and he moved back to High Point.

We are now back living together. Jeff works about an hour away, and I'm still at Guilford. My oldest went to college to study music performance in oboe, but she's taken a bit of time off from school. I think she had it especially tough when her dad moved across the country and we have some things to work through. My youngest is a junior in high school and dually enrolled in high school and college through a partnership program with our local school system and Guilford College. I love having her close to me at work. She is involved in issues surrounding social justice and she's starting to look at colleges. It's an exciting time for her and for me too!

I am thankful to be working at an institution that values undergraduate education. I cannot imagine having to put research above my undergraduate students. There's something very special about teaching at a small liberal arts college. It's a lot of work and there's not much administrative support, but it's a great job and I have a supportive network of colleagues at Guilford and across the country that I've met through various professional organizations. There have been times that I thought it would be easier to walk away rather than work through the difficulties, but I'm glad I've stuck with it, even through the frustrations. I didn't get tenure until I was in my late forties. I received my positive notice of tenure review while Jeff was in Seattle. It was bittersweet to read the letter, and not have him with me to celebrate. I got tenure, but I did not do it alone. My husband and my daughters have been by my side this whole way, and I cannot imagine doing any of this without them. I couldn't have done it without Jeff, and there was never a time that I didn't imagine being a mom. So, while I got frustrated along the way because I wasn't getting "there" as fast as a lot of other women in chemistry, I have fulfilled two of my greatest ambitions: to be a professor and to be a mom.

For women entering the world of academia, my only advice is to do what you love and make decisions for your life based on your circumstances. There are women in the professoriate who came before me who would consider having a child in graduate school the wrong decision. They would advise women to wait until the postdoc or even to wait until post-tenure to start a family. My obstetrician told me that you cannot plan when you're going to have a baby. It happens when it happens, and I agree. If you wait too long, the biological clock can work against you. I've seen that happen to some of my friends and colleagues. I didn't want to be a mom in my mid to late thirties, and if I had waited until post-tenure to start a family, I doubt I would be a mom at all. My path to tenure was certainly a long one, and I've needed to step back, take a breath, and hope and pray that things would work out along the way. It hasn't been easy taking the long road, but I know I'm lucky that I have the support and love of a great family and that I get to go to work and do what I love every day.

Gail and Rebecca in the lab at Guilford College February 2014. Becky is 17 and a junior at The Early College at Guilford High School. She is taking chemistry at Guilford College

Interview Questions

1. How has deciding to start a family or having a family influenced your career? How has your career influenced your family?

Being married to a Ph.D. biochemist and having two children limited locations where I was willing to work. I didn't want to have a long commute, so I was willing to take positions that could have been career limiting (like teaching in a high school). I made the best of the situations, learned what I could at every step along the way, and when I started my tenure track position, I had far more experience than most, which ended up being very helpful.

My career has made my family overly aware of issues surrounding gender equity in the workplace. I've shown my family the data that shows that women faculty at my institution have salaries lower than their male counterparts. My daughters will go into the workforce with their eyes far more open than mine were. I think that when my husband had to deal with my lack of maternity benefits at NC State along with the negative words and actions by my supervisor, it made him keenly aware of the difficulties women scientists face in the workforce. As a group leader and manager of research scientists, he makes sure that the women he supervises receive positive support from him during their pregnancies and that they feel comfortable when issues arise with raising small children. My daughters see the long hours I've had to work over the years. They get pretty annoyed when folks outside of academia make remarks insinuating that I have less than a forty-hour work week. They've dealt with the nights and weekends I've spent grading, planning, going in to work to get projects finished, or traveling for work. I hope they realize that my accomplishments have only been possible because their father has been willing to cook, clean, and do anything else he could to help me along the way.

2. Did you have role models? Which examples were set for you in your childhood or while you were growing up?

My role models were my parents and my teachers. My mom didn't go to work until I was in either fifth or sixth grade. I was the youngest of three, so my brother was already in college and my sister was in high school when she went back to work as a nurse. My dad worked two jobs my whole life. They were both incredibly hard working and made sure my siblings and I had everything we needed. I had some terrific teachers in high school who went above and beyond to make school interesting, fun, and challenging.

3. Have you come up against any significant obstacles during your career and how did you overcome these?

My biggest obstacle was the continual year-to-year contracts as a non-tenure-track faculty member and the feeling of being marginalized at work. It's hard to ask for maternity leave (which was nonexistent for me) when you're worried that your contract might not be renewed if you ask for too much. It's also hard to see others get hired in as tenure-track assistant professors in other departments and progress in their discipline, while you're doing the same work and staying at the visiting assistant professor level, with no voice on campus. I made sure I let my department chair know that I wanted more responsibility and that I wanted to be tenure track. This served me well, because I was given advisees (not usually done), I offered to serve on college-wide committees and was placed on a high-profile committee while still a contingent faculty member. I was able to position myself well when it came to applying for the tenure-track job that opened in my department.

4. Is there anything you would have done differently or would not do again?

Wow. Tough question. I would have worked more efficiently in graduate school. I spent way too much time teaching, tutoring, being a supplemental instruction leader (and then running the program at NC State), and a host of other things rather than staying focused on getting my research finished. In retrospect, I would have done the Ph.D. straight away without stopping for the M.S. and saved a few years too.

5. What advice would you give to young women hoping to pursue a career in academia? E.g., while studying, when planning a family

My advice is to do what you love, do what's right for you, and have confidence. I had two miscarriages along the way and we've dealt with three bouts of layoffs for my husband. It's been tough on our family, but working at a small liberal arts college is invigorating. I have great colleagues who are willing to let me experiment in my classroom and in the lab. I cannot give advice to women who want to become professors at large research institutions, but my path has been right for me. Each institution where I've worked has taught me something about myself and about academia. It is easy to get frustrated when things aren't going as you think they should, but everyone deals with times that are difficult.

Main Steps in Gail's Career

Education and Professional Career

1987	B.S. Chemistry Education, Virginia Commonwealth University, VA
1994	Ph.D. Chemistry, North Carolina State University, NC
1994–1996	Chemistry Instructor, North Carolina School of Science and Mathematics, Durham, NC
1996–1998	Lecturer and Visiting Assistant Professor, North Carolina State University, NC
1999–2000	Assistant Professor, University of the Sciences in Philadelphia, PA
2003–2007	Visiting Assistant Professor, Guilford College, NC
2007–2010	Assistant Professor, Guilford College, NC
2010–present	Associate Professor, Guilford College, NC

I Finally Know What I Want to Be When I Grow Up

Catherine O. Welder

Photo: Amy Chan Photography

Why is it that we are always expected to have an answer for what we want to be when we grow up? From a very young age children are asked what they would like to become. Fireman, teacher, or my youngest son's current choice, a monster truck driver. It starts with preschoolers. Then, as we hit the teenage years, career aspirations become more of a threat. My husband heard this at one point in his teenage years: "You don't want to dig ditches when you grow up, do you?" We face serious choices as we complete high school, and for those of us who go on to college, we are hit with another round of more ominous decisions a few years later. What, exactly, do we plan to do after graduation? A few people are fortunate to have a clear picture of where they are going next. Others seem to avoid the decision-making process as long as possible or take some time off, perhaps a gap year, as they decide where to go next. But the underlying question is always the same. What do I want to be when I grow up?

It wasn't until well into my grad school days that I discovered a love for teaching. I served as a teaching assistant for general and organic chemistry courses all the way through my graduate student years, all 5+ of them. While most of my responsibilities involved helping students navigate the labs, I also had one-on-one

---

C.O. Welder (✉)
6128 Burke Laboratory, Dartmouth College, Hanover, NH 03755, USA
e-mail: Catherine.O.Welder@Dartmouth.edu

contact with students who needed additional help with the lecture portion of the course. It was during those office hours that I discovered I truly enjoyed helping people better understand complicated topics. I got a real thrill when a student finally had an "Ah HA!" moment.

I still remember the day I told my Ph.D. advisor that I wanted to pursue a career in academics. I was in my 3rd year of the program. His response was simple. "Good. I don't think you would like industry." While I have never held an industrial position, I think he was correct. I love teaching!

My dad was an industrial organic chemist. He served as the Director of Research at his firm. He was very excited when I took geometry in tenth grade and did well. "Great!" he said. "You will do well in chemistry." I didn't understand how he could jump to that conclusion. Now I see what he meant. The concepts you master when solving proofs in geometry are similar to the skills needed to master syntheses in organic chemistry. And being able to visualize in 3D starts with geometry and is critical in organic. I went on to take chemistry in both my junior and senior years of high school and headed off to college with the conviction that I had found my major.

When I got to Wake Forest University, I could choose between a Bachelor of Arts (B.A.) or Bachelor of Science (B.S.) degree in chemistry. The tracks were the same for the first two years, so I didn't have to decide right away. Once it was necessary to make a choice, I noticed that the B.S. would be more work. Who wants more work? "I'll just get a B.A.," I thought. Well, when my dad caught wind of that, he told me to earn a B.S. in chemistry or pick another major! He assumed I would have difficulty finding a job with only a B.A. He was probably right. The people choosing B.A.s in chemistry were mostly on a premed track, which was of no interest to me. So, B.S. it was.

Amongst all this, Frank popped the big question. Would I marry him? We began dating as seniors in high school and went to separate colleges, seeing each other about once a month. This was back in the day before Skype, cell phones, and even e-mail. A month was an excruciatingly long period of time between visits. If we could make it through that, we could make it through anything, so I said yes. That was during the fall term of my senior year, the same term that I was applying to various graduate programs in chemistry. My dad was convinced that if I married, I wouldn't pursue a graduate degree. Or perhaps I would start and not finish. I was convinced that if I had to choose graduate school *or* Frank, I would choose Frank, but there wasn't a need for this to be an either/or situation. We had been dating for five years, which seemed like an eternity, when we married. I needed his strength and companionship to have the endurance to survive a Ph.D. program. Why not be married *and* be a graduate student?

As I began visiting chemistry departments in the spring semester of my senior year, I was thinking about the possibility of stopping at a master's degree. Again, it was about the work. A Ph.D. program is so much more demanding, and I wasn't sure I had it in me. That's when my dad gave me another important tidbit of advice. "Don't apply to M.S. programs. You can always enter a Ph.D. program and stop at a master's degree." Again, he was right. Graduate schools only have so many spots in each year's entering class. If the school has a choice between accepting a student who wishes to pursue a master's degree and one who wishes to pursue a doctorate, the doctoral student often receives the spot. I now know that many who begin a Ph. D. program in chemistry don't actually complete it. I was told by a more senior graduate student during orientation to look to my right and to my left. Two of the three of us would not complete our Ph.D.s. I literally thought, "Where are *they* going?" The prediction held true. About a third of my entering class completed Ph. D.s in that program. However, most students left by choice and schools vary drastically in attrition rates.

Frank and I married as soon as we had both completed our undergraduate degrees. I chose the Georgia Institute of Technology for graduate school and, thankfully, Frank found a position as a chemist in an environmental lab nearby. Though we always wanted to have kids, we didn't want them right away. School was demanding and time-consuming. So was marriage. I remember our first year of marriage as particularly tough. We had discussed divorce in premarital counseling and had agreed that it would not be an option for us. I can't tell you how much freedom that gives us to be ourselves, knowing that despite the current situation or argument, the other will always be there.

Graduate school was certainly a challenge, but it was never overwhelming. (Then again, it's been a number of years since I was in graduate school, and perhaps I've blocked the worst of it from my memory.) I did well in my coursework and enjoyed my area of research. But I should mention a nugget of advice from my undergraduate research advisor regarding grad school selection. He advised me to select a school that had multiple research groups that I would consider joining. Excellent advice! You never know if a particular professor might be considering retirement, a position at another school, or if he or she is not accepting students due to the current size of the research group or the funding situation. I chose an advisor who worked in the area of organometallics. Historically, his group was half inorganic students and half organic students. I chose organic chemistry as my primary field of study only after another talk with my dad, during which he informed me that I would never find a job as an inorganic chemist. While there are inorganic jobs out there, it does seem that it's easier to find a teaching position as an organic chemist.

My biggest challenge in graduate school came in the form of finding balance, both within the program and outside of it. Within the program I needed to balance the time it took to study for classes and cumulative exams, to serve as a teaching assistant, and to make progress toward my research. At some point my research advisor questioned whether or not I was putting enough time into my research. I spoke with my husband and determined for myself that I was giving all I was willing and able to give. Either it was enough to successfully complete the program, or it wasn't. I couldn't give more. I needed to balance the daily grind of graduate school with extracurricular activities, too. I was very active in the church choir, handbell choir, and even the softball team. I needed the fellowship, friendship, and exercise. I needed the outlets and the support system, or I would not have successfully completed my Ph.D. Balance was difficult to achieve.

One night, when my advisor was hosting his research group for dinner, his wife asked me what I hoped to do after graduate school. My advisor overheard the question. They had raised seven children of their own, and she had stayed home with the kids. I mentioned that I was thinking about being a stay-at-home mom, especially while my kids were young, and that I was considering homeschooling. The concept of obtaining a Ph.D. degree and then being a stay-at-home mom was so foreign to my advisor that he literally didn't understand what I had said. He could not fathom earning a Ph.D. and then not using it to launch into a wonderful, long career in chemistry. For his own clarification, he restated what he thought he heard, which wasn't even close to what I said. I did not correct him, and the topic never came up again. His reaction was very interesting to me. I remember it well as a young woman trying to figure out how pursuing a graduate degree in chemistry would fit in with raising a family. I did not see education as a waste of time or conflicting with having children and perhaps even homeschooling them. To me, it was again about balance.

Finally, after 5½ years in graduate school, I successfully defended my Ph.D. thesis. It likely would have taken me even longer if two former group members had not remained in the department, offering me assistance as needed. The two of them served as a tremendous resource after my advisor retired a full year before I completed my degree. As his last student, I was alone in the lab for my final year. I feel that no matter how long it takes you to complete a Ph.D., that last year is the most significant in terms of data collection. Bitterness can set in as the degree seems so far away, yet you have invested too much to stop now. It had been a long road, and I was glad to be at the end of it.

I had not thought much about what to do after graduate school. Frank wasn't happy in his job and was ready for a change, but he wasn't actively looking for any positions. I am thankful that I was able to find a short-term postdoctoral fellowship within the chemistry department at Georgia Tech. My new advisor was convinced that he would have more funding soon but did mention when he hired me that he only had funding for a 3-month appointment. I would be carrying out synthetic work on an NIH-funded cancer research project. Just after I was hired, the government shut down. That froze additional funding, and after 3 months my position was terminated. My advisor was devastated by the news and extremely apologetic. I, on the other hand, was slightly relieved. I had discovered a wonderful lesson. I don't like synthesis! It's tedious, slow, and I found it a bit boring. I was glad to move on to something else.

It turns out the "something else" was another postdoctoral position on the Georgia Tech campus within the Institute of Paper Science and Technology, a school designed to train graduate students in the science and engineering of papermaking. I would be studying ozone bleaching of wood pulp, a mechanistic project. After six months or so of pure research, I was given the opportunity to cut back on research hours and teach a survey organic chemistry course to the first-year graduate students in the paper science program. It was a Godsend! I taught the 10-week survey lecture course that covered all the basics to prepare students for advanced courses in pulping and bleaching. I gained teaching experience at the collegiate level, a tremendous addition to my resumé.

As I completed my first year as a postdoctoral fellow, my husband and I began to seriously consider our next career move. He received an offer for a chemistry position at the bachelor's level in Louisiana. In the same time frame, I received an offer from my alma mater for a one year, nonrenewable teaching position. If we moved to Louisiana, we would try to settle in and start a family. If we moved back to North Carolina, I would have a temporary job, and Frank would work on his golf game. We weren't sure what to do, and we spent quite a lot of time in prayer and asked friends to pray for us as we made the decision. It almost seemed like a no-brainer to take the full-time, permanent position in Louisiana, where my husband would be the primary breadwinner, and I could be the stay-at-home mom I had been thinking about all those years. But that wasn't God's plan for us. We had one weekend to decide between moving to Louisiana or North Carolina. I had been given an extension for making a decision at Wake Forest University, and Frank had only a verbal offer from his potential employer in Louisiana. We prayed that God would provide specific information to help us to make the best decision. He did. Within a few hours, Frank's written offer arrived with no mention of health benefits for his spouse or family. To me the whole point of accepting his position would be to start a family, yet dependent medical expenses wouldn't be covered? We considered this flaw to be the answer. Monday morning we accepted the position at Wake Forest.

A few months later we arrived in a new role in a familiar city. I believe it was my second day in the office when a senior colleague approached me asking what Frank was doing. This professor wanted to hire Frank to set up, learn to use, and teach his research group to use a piece of equipment that Frank had never used.

Me: "You understand he's never worked on this type of equipment, right?"
Professor: "Right."
Me: "Yet you wish to make him an offer?"
Professor: "Yes. Well, I should probably talk to him first. Have him call me."

Frank started working for him that week. We hadn't finished unpacking, and Frank hadn't picked up a golf club.

Frank had great success working in the research lab. As a matter of fact, his boss was also the chair of the committee that determined who would be accepted into the chemistry graduate program. Once we learned that I would be at Wake Forest for the fall term, Frank applied to the graduate program in the chemistry department, but he was denied admission, likely because it was so late in the application process and all the slots had been filled by other applicants. His boss let him know that his application would be accepted for spring term (January) admission. No one had dropped out of the program. Why was there suddenly a spot available to him? (We view it as another Godsend.) So, Frank launched into his graduate studies only months after we moved to NC. He began graduate school during my first one-year, nonrenewable contract. For each of the next five years, I received one-year, nonrenewable contracts. Frank was able to complete not only his Ph.D. in analytical chemistry, but also a joint M.B.A. from Wake Forest's business school while I taught.

Wake Forest, like many U.S. colleges and universities, makes tenure decisions after six years of service. My position was not on the tenure track, so after teaching for six years, I knew that I would not be able to stay for a seventh. The department could not figure out how to extend my stay as a non-tenured faculty member. Frank would finish both degree programs, so we began asking the all-important question again: what do we want to be when we grow up?

By this point we had been married 13 years, and we still didn't have kids. Starting a family became more important and even somewhat urgent. As such, we decided that Frank would look for full-time employment with the hopes that I would be home with a baby within a year or so. Frank was able to find a job in Louisville, KY. While it wasn't his dream job, he was more than qualified and accepted the position.

Frank and Cathy in the Lost River Gorge and Boulder Caves, North Woodstock, NH

Soon after moving to Kentucky, I discovered I was pregnant with Luke. There were some complications during the pregnancy, and I was even hospitalized for ten days when I was 11 weeks pregnant. After being released I was "allowed" to visit one doctor or another two to three times per week for the duration of the pregnancy and even beyond it. I don't know what I would have done if I were employed! For me, being pregnant with complications was a full-time job. I think that it was another Godsend that we were in Louisville with an OB-GYN and consulting specialists who had much experience with cases such as mine and knew exactly what to do.

Frank was dissatisfied with his job, and we missed friends and family terribly while in Kentucky. Once I received medical clearance, Frank began looking for a new job. He found a position that allowed him to work almost exclusively from home, and we lived in Kentucky for almost a year while trying to sell the house in preparation to move back home, North Carolina. In the meantime, friends from

Wake Forest heard that we might be moving back to the area, and they contacted me to see if I might be interested in teaching there again. (Yes, you guessed it: one-year, nonrenewable contract.) I agreed and we moved back to the Winston-Salem area.

During my first year back, I became pregnant with our second child, Ethan. He was due in late September, and I faced another one year contract, so I had a decision to make. Did I want to start the fall term eight months pregnant, take six weeks off for maternity leave, and then finish the fall term with a very young child in day care? That was one of the easiest decisions I've ever made. NO! I knew from my first pregnancy that there would be many doctors' appointments in my future, especially in the third trimester. I also knew that at six weeks babies aren't sleeping through the night, which means parents aren't sleeping through the night. And I firmly believed that I could take better care of my six week old than could even the best of day cares.

When deciding how much time to take off, I considered my own interests, as well as those of my students, coworkers, and replacement. I did not think it would be in the best interest of students to have me for the beginning and end of the 15-week term and a replacement while I was out on maternity leave. I knew a 6-week position might be difficult to fill and would not be appealing to strong applicants. So that the department would have plenty of time to carry out a search, I informed our chair during my first trimester that I planned to take the next year off. If I had it to do over, I would probably elect to wait until later in the pregnancy to announce my intentions. However, it all worked out in the end. I was able to stay home with Ethan for almost a year before accepting my final one-year, nonrenewable contract (the eighth) at Wake Forest.

During my eighth year at Wake, it became evident that I would always have the uncertainty and corresponding rank that a one-year, nonrenewable contract brings with it. In the spring of that year, I reluctantly began looking at other schools. The department chair was well aware of my situation and asked me if I had seen a late ad in *Chemical and Engineering News* (C&EN), a weekly publication of the American Chemical Society, posted by Dartmouth College. I had not. They were looking for someone to direct the organic lab courses year round and to teach the summer Organic II lecture course. I had exactly the qualifications they listed in their advertisement. Most schools advertise chemistry positions early in the fall term, arrange November/December interviews, and then make offers for the following summer or fall appointments. Dartmouth placed the ad in February or March and held interviews through the end of April hoping for the candidate to start in June. If they had placed their ad the previous fall, I would not have seen it and would not have been considered for the position. God's providence was evident to us again.

So, here I am at Dartmouth. This is exactly what I want to be when I grow up—a full-time teacher and mentor. I'm not required to write grants, though I'm certainly not discouraged from doing so. I'm not expected to carry out research, though any type of scholarly activity is a plus. I'm not facing a tenure decision, and I consider it a step in the right direction to have three-year, renewable contracts. My official title is "Senior Lecturer." While not many schools have lecturer positions, they are becoming more popular and some schools are creating tenure-track teaching positions.

I teach year round, but some schools have nine- and ten-month positions. I would prefer having summers off to enjoy with my children, now 5½ and 8. Our organic labs are six hours long and begin at 2 pm, which means there are several nights in which I get home after my kids are in bed. But my colleagues were willing to help me address the concern of rarely being home after the kids' school day. As I negotiated the details of my offer, I noted that labs typically run four nights a week at Dartmouth during the Fall and Winter quarters and two nights a week during the Spring and Summer quarters. I asked if it would be possible for me to stay two nights a week each term and for other organic faculty to stay for additional lab periods. I never thought they would agree, but they did! Labs typically meet for eight weeks each term, so that's only 32 weeks a year that I have what we've come to call "Late Nights." That's manageable. The kids completely understand that Late Nights typically mean Mom won't be home until after bedtime, though I must admit that I'm not too upset when students complete their experiments early enough for me to give my kids a kiss and tuck them in bed.

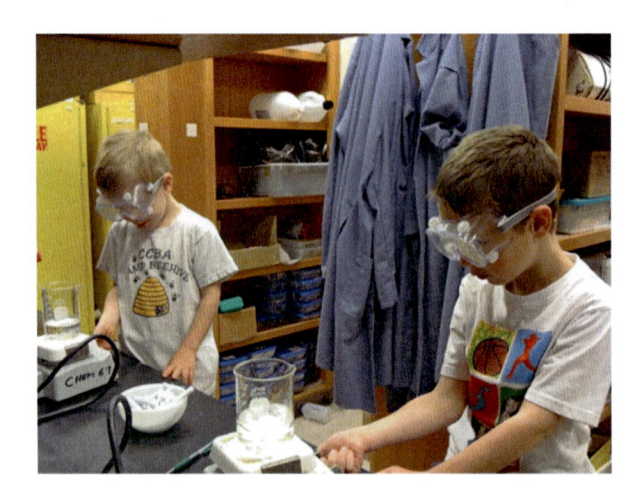

Ethan and Luke explore with magnetic stir bars and a stirring hotplate (and water, of course!)

Ethan and Luke boating in Virginia

For the most part, Frank is a stay-at-home dad. He shuttles the kids back and forth to various activities and is available as needed. Since my job requires me to be at work much of the afternoon and evening hours, Frank is limited in his job options. He must be available to pick up the kids no later than 5:30 pm, the typical closing time for day cares and after school programs in our area. Last year Frank accepted a one year teaching position for general chemistry lecture and lab at a 4-year college located about forty minutes from our home. That was his first teaching experience, other than serving as a teaching assistant in graduate school. He enjoyed it. The positive experience was enough to encourage him to apply to teach a first-term general chemistry lecture course at Dartmouth College. He just finished that course and plans to teach at Dartmouth again in future terms. The timing works well as lecture courses are held in the mornings and my labs meet in the afternoons. That means that if the kids are out of school, one of us can be available to watch them without interrupting our work schedules. As Ethan starts kindergarten in the fall, I think Frank will be even more interested in intellectual pursuits. He has played a very crucial role in childcare for the past 3½ years, and I am grateful that one of us has been home for most of our children's preschool years.

So, for us, the balance of teaching and family has been rather successful. I feel that a career in academics, with a somewhat flexible teaching schedule and the possibility of having summers off, is much more attractive than a career in industry. Balancing two careers can be difficult, but Frank and I have managed each career choice so that one of us has the primary breadwinning role while the other works and manages things on the home front. We have never felt that the pursuit of one career has come at the expense of the other, but I think that is because one of us has had an obvious career-advancing choice at a time when the other's job was stagnant (Frank) or temporary (me).

As you begin making career choices, take comfort in the fact that there are often multiple pathways to success and happiness. Those who pursue and complete doctorate degrees often continue in a lifetime of learning. One of my most valuable lessons in graduate school was that I can now teach myself. I can read books and papers to learn more about a topic of interest. I can extend my knowledge by checking the references, and I can contact the authors for clarification. Choosing a career field does not necessarily limit future options. Instead, your degree and first job set you up for a lifetime of exploration. Collaboration with colleagues is becoming vital. You can pair your area of expertise with that of another and venture into fields currently beyond your imagination. I wish you the best as you find your niche, whether it's as a student, as a postdoctoral fellow, in a first job, or as a result of a career path change.

Welder family photo. Photo: Amy Chan Photography

Interview with the Author

1. Why did you choose chemistry in the first place?
For me, chemistry comes easily. I struggled to do well in some of my humanities courses and don't particularly enjoy reading. Since my dad was a chemist, he was obviously pleased with my choice. Plus, not many parents complain when their kids pick a major that tends to lead directly to a field of employment. I guess it comes down to the fact that I did well in my courses and enjoyed them.

2. How has deciding to start a family or having a family influenced your career? How has your career influenced your family?
My husband and I postponed having children until after we both finished graduate school. We married right after completing our undergraduate degrees. I

completed my Ph.D. in organic chemistry followed by a postdoctoral fellowship while he worked. Then, I worked full-time in academics while he completed joint degrees, a Ph.D. in analytical chemistry and an M.B.A. So, we were married 14 years before our first son was born.

My husband and I seem to take turns at being the primary breadwinner. The person with the secondary career has served as primary parent and keeper of the home front. He worked full-time when I was in school and then the roles reversed. We elected to follow his career once he finished school and to start a family. He worked full-time when our first son was born, and I was able to stay at home for about 2 years during the pregnancy and early months. Then, my husband began working from home (flexible hours!) when I went back to teaching full-time. I took one academic year off when our second son was born. Alternating roles has worked well for us.

I have a colleague who agrees that it's tough to balance two demanding careers with family. It seems to work well when one person has flexible hours or part-time employment for the inevitable sick days or taxi service aspects of extracurricular activities.

3. Did you have role models? Which examples were set for you in your childhood or while you were growing up?

My dad was my role model growing up. I'm not sure how he did it, but he managed to balance a career directing an industrial research lab with a hobby that blossomed into a full business AND to attend all my sporting events. My dad was also a deacon in the church and active with the youth group before I was old enough to be a part of it. Looking back, I realize that he must have made me a priority as I was a member of one sports team or another virtually year round, and he attended not only the home games but the away games as well, many of which had an early afternoon start time.

4. Your father seems to have had quite an influence on your career path. Do you feel that he pushed you in a direction you didn't want to go?

No, I never felt uncomfortable with my dad's comments and influence. While his comments seem blunt, and perhaps harsh, I never felt that he was pushing me in a direction that I didn't want to go. He was simply helping me make decisions that fit my strengths and wishes.

5. Have you come up against any significant obstacles during your career and how did you overcome these?

Two things come to mind. The first is a conflict I had with my Ph.D. advisor. He was from the era of an all-male research group who worked 6 days a week, perhaps ten hours a day. Somewhat by choice and somewhat by necessity, I did not work that many hours a week. I was the last of my advisor's graduate students, and he retired a full year before I defended my thesis. I carried out most of my research in an inert atmosphere glove box due to the extreme reactivity of the reagents I was using. As such, I refused to work alone in the lab, and to this day I feel that was the correct decision to make. I could only work in the lab when others were working in nearby labs. But, I was also balancing the demands of graduate school with those of being newly married. That first year of marriage, in particular, takes a great deal of work if the relationship is going to be successful for the long term. I gave all I could give toward completing a Ph.D. and was fully aware that others may deem it insufficient. I was satisfied that I was doing my best to meet the various demands in my life. If it wasn't good enough, then I just wouldn't earn my Ph.D. Once I completed several of the degree requirements (such as courses, seminars, and cumulative exams) and could focus more on research, things moved along nicely.

The other obstacle I have faced is the fact that few schools, though more now, have lecturing positions. I am not, nor have I been, interested in a traditional tenure-track position. I do not have an interest in carrying out research. Fortunately, I have been blessed to find positions without much difficulty at schools that do support the full-time lecturer.

6. Is it common for academic institutions to make accommodations for working moms like Dartmouth has made for you in regard to covering late lab nights?

I'm not sure what other colleges and universities do to accommodate working parents. I think it is becoming more common for families to balance two jobs, and as such employers are creating more options for their employees, such as flexible hours or the ability to work from home. I have found departments to be sensitive to the needs of families. I do think that this is a topic for discussion with a potential employer after you have received an offer and as you negotiate the details of the position.

7. Tell us more about Luke and Ethan.

Luke (8) is my little scientist/engineer. He enjoys math and truly understands the fundamental concepts he's seen thus far. Of course, I use every opportunity to stretch his math skills in everyday life, which currently means exploring multiplication, division, fractions, time, and money. His favorite TV show is How It's Made, though he enjoys watching Power Rangers as well. He likes alpine skiing in the winter and biking in the summer. We try a little bit of everything including baseball, tennis, soccer, and swimming. He's tenderhearted and will probably choose a few people to be close friends instead of many acquaintances. Luke prefers quiet, reflective environments. He has thoughtful conversations and still allows cuddles and hugs.

It's amazing how two children from the same gene pool can be so different. Ethan (5 ½) is loud and full of energy! He does his best to keep up with his big brother. He has never met a stranger and even gave one man a complete family history in a short elevator ride. He always runs to the door when I get home to give me a big hug and yells, "Mommy's home" over and over again until he gets to me. Ethan loves everything active and is quite coordinated. His favorite activities include alpine skiing and gymnastics. He also enjoys soccer, biking, swimming, hiking, baseball, tennis, and playing on the playground. He likes looking at books and can't wait to learn how to read. He is very caring and considerate of others and will even share items from his candy bag. When a preschool classmate missed a day of school, even though crafts aren't his favorite, Ethan completed two crafts that day so his friend would have one as well.

8. How does Frank feel about being a stay-at-home dad?

He says it's the hardest job he's ever had. In the working world, we accomplish goals and are rewarded financially for our efforts. Recognition for a job well done makes us feel good. At home, there are fewer short-term goals, so the sense of accomplishment is less frequent. Children are less likely to reward a job well done with praise and admiration. They are, by nature, needy and immature. However, we feel that having a parent at home during these formative years is worth the sacrifice and effort. Frank says that he finds other stay-at-home dads, who understand his perspective and struggles, a great source of support.

9. Is there anything you would have done differently or would not do again?

I'm not sure that I would have done it differently, but waiting to have children has its pros and cons. A pro of waiting is that both my husband and I were able to independently complete graduate degrees without having to factor children into the mix. But, waiting to have children does have its cons. It's harder to become pregnant, remain pregnant, and to deliver a healthy baby as you age. Second, your peers likely had children years ago, meaning you can hang out with the 20 somethings who have children the age of yours, or you can interact socially with people your age who have children in middle and high school, perhaps even college when your children are in preschool and elementary grades. Again, I don't think I would change anything, but this is certainly something to carefully consider when younger. In fairness, though, there does seem to be a trend among highly educated women to postpone having children, so in the workplace there isn't such a gap between the age others began having children and when we did.

10. Has it been difficult to make your various career decisions, especially when considering your spouse's career, too?

Yes, each career decision was difficult at the time. I think it is rare to have job offers for both partners when a move is involved. Trust in and open communication with your partner are vital. I followed Frank to Kentucky hoping to either become pregnant and a stay-at-home, or to find a collegiate teaching position. Frank followed me to New Hampshire and still hasn't found a full-time job, but he has certainly strengthened his resumé in a new (teaching) career by teaching part-time at Dartmouth College.

11. You mentioned that one reason you chose Georgia Tech for graduate school was that a number of faculty members were carrying out research that appealed to you. Were there other significant factors that influenced your choice of graduate programs?

Yes, there were several factors that ultimately led to my decision to attend Georgia Tech. I was only considering programs in the southeast. I did not want to be more than a 6-hour drive from home (western North Carolina) as I figured I would be traveling frequently during my first term of graduate school to visit my fiancé and to plan our wedding. That seems like a rather petty reason now, and I did not travel as much as I thought I might. More significantly, though, I was offered an additional $4,000 stipend during my first year at Georgia Tech. Since the typical stipend at the time was only $12,000, the additional $4,000 was substantial.

If I were advising students what to consider when selecting a program today, I would encourage them to consider various aspects of the degree program. How many classes are you expected to take? Is there a breadth requirement? (I had to be proficient in 3 of the 5 areas of chemistry—analytical, organic, inorganic, physical, and biochemistry. Proficiency was achieved by either a high score on a standardized entrance exam or by taking classes in that area.) How often will you be asked to serve as a teaching assistant? (At many schools that often depends on whether or not your research advisor has external funding.) What are the additional requirements, such as cumulative exams, seminar presentations, and research proposals? You might even consider how long it takes the typical candidate to complete the degree program. Visit the school and talk with current students in the program, if you can. Of course, financials often play a factor as well. Are you selecting a field of study, such as chemistry, that supports its students with stipends? What is the cost of living in the area you are selecting? I believe several of the schools I was considering would have helped me meet my career goals.

12. **What advice would you give to young women hoping to pursue a career in academia? E.g., while studying, when planning a family**

I recently read from another female author that there will never be a good time in your career to have children. She and her husband elected to start their family while she was still in graduate school. She felt that interviewing for postdoctoral positions while 8 months pregnant put her at a disadvantage, as interviewers wondered how committed to the job she would be with a young child at home. Discrimination is worse in some workplaces than others, but I feel that it may be a common first reaction from those in the workplace who hear of a colleague's pregnancy to be, "Great. What are we going to do now?" as if some tragedy just struck. Women have been balancing career and family for years. We can do it. I feel that academia offers women the most flexible job options regarding family. You may be awarded an extra year before facing a tenure decision, for example. I agree that having children typically doesn't lead to immediate career advancement. However, advancing your career is not likely going to be the only item on your lifetime achievement agenda. Family is important. Relationships are vital. Start your family when it feels right for you.

Main Steps in Cathy's Career

Education and Professional Career

1990	B.S. Cum Laude with Honors in Chemistry, Wake Forest University, NC
1996	Ph.D. Organic Chemistry, Georgia Institute of Technology, GA

1996–1998	Postdoctoral Fellow, Institute of Paper Science and Technology, GA
1998–2004	Visiting Assistant Professor, Wake Forest University, NC
2006–2007	Visiting Associate Professor, Wake Forest University, NC
2008–2009	Visiting Senior Lecturer, Wake Forest University, NC
2009–present	Senior Lecturer, Dartmouth College, NH

Honors and Awards

2013 Dean of Faculty Teaching Award for Visiting and Adjunct Faculty, Dartmouth College

Over the last 20 years, Cathy has taught a variety of undergraduate and graduate chemistry courses. Her academic interests include demystifying organic chemistry, developing new laboratory experiments, and training student teaching assistants as the next generation of educators.

Acknowledgments I would like to thank three of my former Dartmouth students for reviewing this chapter and providing valuable commentary: Rebecca E. Glover ('11), Alexandra T. Geanacopoulos ('13), and Marissa H. Lynn ('13). All three are now considering graduate programs in the sciences and they represent our book's target audience. Their feedback has been invaluable.

My Not-So-Secret Double-Life as a Chemistry Professor and Mom

Kimberly A. Woznack

I'm leading a double-life as a Chemistry Professor and as a Mother. My double-life is, however, not a secret. My university colleagues all know that I am also a mother. My family is also very aware that I am a Chemistry Professor. As I write this chapter, I've been a chemist for the last 17 years, a professor for the last 10 years, and a mother for the last 6 years. While I cannot claim to be the perfect role model for those considering a similar double-life, I can say that I am very happy doing both. I cannot imagine my life without either of these two parts.

Childhood

I grew up in upstate New York. My dad is an Optometrist, and while my two brothers and I were very young, my mom stayed at home with us. When my dad began his own private practice, my mom transitioned into being his full-time Office Manager, and she made her own schedule. (This schedule flexibility is very useful when it comes to being a parent. While the class schedule of a faculty member

---

K.A. Woznack (✉)
Department of Chemistry & Physics, California University of Pennsylvania, 250 University Ave, California, PA 15419, USA
e-mail: woznack@calu.edu

doesn't allow for infinite flexibility, I am often glad that I can schedule appointments for my children, during regular business hours, at a time when I don't have classes or office hours.) My parents always prioritized our education and gave us educational toys to play with. In both elementary and middle schools, I was part of some accelerated classes, and my involvement in advanced science courses set me on a path toward academic success.

Guilderland High School, Training Future Scientists (1989–1993)

In high school, I decided I wanted to become a Biologist. I was fascinated by human anatomy. My high school chemistry experience wasn't as exciting. I was enrolled in the "research" chemistry course and we designed our own experiment and wrote up a paper about it. In order to accomplish this many regular experiments were cut. So, my love for chemistry did not yet to come to fruition. I was not even aware at the time how fortunate I was to grow up in an upper-middle class environment with an amazing school district. Many of my best friends from high school went on to obtain graduate degrees in science. To this day, I am grateful to have benefitted from attending a school, which offered so many opportunities. During high school, my youngest brother was still a toddler. I have to say that watching how much work it was for my parents to balance their work and home lives made me think that I wasn't interested in having children for a very long time.

A Partner for Life (1992–Current)

During high school, I also met the love of my life, who is now my husband, Ray. We had mutual friends, similar taste in music, and our fathers both liked to shop at Campmor, an outdoor goods store in Paramus, NJ. We both had the Campmor T-shirts to prove it. We were never in the same classes, but we enjoyed spending time together. Ray went to a local community college after his graduation for their automotive program and would even drive me to school my senior year.

Kim and Ray at post-Prom Party (1992)

College Professors as Mentors (1993–1997)

I attended Hartwick College, in Oneonta, NY. I was fortunate to have been awarded the Abraham Kellogg Merit Scholarship, worth full tuition for my first year and renewable until graduation. Hartwick was only an hour away from home and many of my friends from Guilderland also enrolled at Hartwick.

I studied General Chemistry I and II diligently. Chemistry was a far more logical science than it had appeared in high school. Studying with my lab partner for the chemistry exams really helped me see how much I liked teaching. She was really struggling with the material, and I wasn't. When I explained it to her, it reinforced my own understanding. I earned the CRC Freshmen Chemistry Award and my own copy of the CRC Handbook. My interest in chemistry was piqued. Meanwhile, my interest in biology was beginning to wane. I took courses in Genetics, Molecular Biology, Ecology and Microbiology. With the exception of perhaps Developmental Biology, these courses didn't seem to excite me.

For a while, I was a double major in biology and chemistry and I took the courses needed for both. Dr. Bill Vining, who was my chemistry advisor, had an enthusiasm for chemistry that was contagious. In my junior year, I was able to attend the Spring 1996 American Chemical Society (ACS) National Meeting with the Hartwick College Chemistry Club. Several of my friends were presenting posters of their research. I will never forget walking into the New Orleans Convention Center at the first meeting and realizing that all of these people were chemists. I didn't know that many chemists previously, and to look through the program and see all these different names, I was floored.

I was encouraged to do a Summer Research Experience for Undergraduates (REU) program and found this to be really interesting. I participated in the National Science Foundation (NSF) Solid State Chemistry REU program. This program began with a weeklong workshop on solid-state chemistry at SUNY-Binghamton and then participants dispersed around the country to complete projects. At the end of the 10-week experience, the participants returned to SUNY-Binghamton to give an oral presentation summarizing their work. My summer work was in the Department of Materials Science and Engineering at Cornell University. Completing the REU solidified my desire to go to graduate school in chemistry. With the dual major in both biology and chemistry, I was only able to earn a B. A. in Chemistry. If I dropped the biology courses my senior year then I would be able to complete a B. S. in chemistry and enhance my chances of getting into graduate school. It was decided, I was going to be a chemist.

In my senior year, Dr. Susan Young started her tenure-track position as Hartwick's Inorganic chemist after Bill left for a position at UMass-Amherst. I completed my senior research project under her supervision. I presented my results in a poster format during the 1997 ACS National Meeting in San Francisco. While at that ACS meeting, I was able to get Chemist Glenn Seaborg's autograph on a T-shirt that says, "I'm in my Element." To this day I have the signed T-shirt, and a picture of myself getting it signed, framed, and mounted in my office. I don't know if I will ever meet another Nobel Prize winner, but I love to share this story with my students.

Kim meeting Glenn T. Seaborg at an ACS National Meeting (1997)

I completed my senior research project under Susan's supervision and spent a lot of time in her office. This is when I got a glimpse into what the life of a faculty member is really like. She would not only have classes to prepare for, teach, and grade papers for, but I saw the committee work she was involved with and the meetings she had outside of her office. I remember sitting in her office one day and saying, "I could do this. I think I would like to be a chemistry professor." I couldn't imagine what the daily life of an industrial chemist would be like. I am not confident that I would have envisioned myself so clearly in the role of "chemistry professor" if Susan had not been a female professor.

During my entire, undergraduate career, Ray and I stayed together as a couple. It was a long-distance (70 miles) relationship, but we made it work. We talked on the phone and we made many drives along Interstate-88 to visit each other on the weekends. (This was before Facebook, Skype, and FaceTime.) I even did my REU within New York State so that I wouldn't have to spend an entire summer outside of driving distance.

As I was finishing my undergraduate degree, the conversation turned to what would happen when I went to graduate school. Ray told me that he would move with me wherever I went to school. He was ready to leave the Capital District of New York State and see another part of the country. It was an enormous relief to me that I could look at schools without worrying about the impact this choice would have on our personal relationship. I applied to and was accepted by many graduate programs.

Kim and Ray, Hartwick Graduation Day (1997)

Taking Chemistry to the Next Level, Moving to the Midwest (1997–2002)

I chose to start the chemistry Ph.D. program at the University of Wisconsin-Madison. Ray and I moved to Madison in August 1997. Ray found a job very shortly after we moved working for a Honda automotive dealership. They had two locations and he would stay with them the entire five years we were in Madison. I began a research project supervised by my chemistry advisor, Dr. Arthur B. Ellis, as well as a chemical engineering professor, Dr. Thomas F. Kuech. This project made use of the Synchrotron Radiation Center (SRC) located fifteen miles away in Stoughton, WI. I am thankful for having been mentored by both Art and Tom during graduate school. They both exhibited different leadership styles, and I incorporate some aspects of each of their styles as I mentor undergraduate researchers. I am really thankful for all of the connections I made to chemical educators through my membership in Art's research group. I met fabulous people from around the country who prioritized many aspects of undergraduate education and introduced me to many ways of teaching inorganic chemistry and topics related to nanotechnology. During graduate school, I only ever had one female faculty member, Dr. Judith Burstyn for any of my graduate courses. While the faculty members at UW were predominantly male they were also predominantly parents. I met the children and families of many of the faculty, so this made it seem achievable to have a family and an academic career.

Ray and I really enjoyed our years in Madison, WI. We made many friends in the area, both chemistry graduate students, and otherwise. Ray's interests and hobbies introduced us to Wisconsin natives with common interests. We enjoyed traveling to go four-wheeling or dirt bike/quad riding in various areas of Wisconsin, the upper peninsula of Michigan, and the sand dunes on Lake Michigan's eastern shore. We went snowmobiling in Yellowstone National Park and in the neighboring areas of Idaho and Montana, twice. At this point, I was nearing the point of becoming a professor but was not yet entertaining the idea of becoming a mother.

While participating in a curriculum workshop at UW, one of the project's evaluators mentioned to me that she knew someone who was looking for a postdoc in the realm of chemical education. She connected me with Dr. Christopher Bauer, of the University of New Hampshire (UNH) who had indeed obtained a grant to implement and study the impact of all of the NSF Chemistry Systemic Initiatives in the UNH General Chemistry Program. This was exactly the opportunity I was looking for. I knew I wanted to teach, at a predominantly undergraduate institution, but I knew I wasn't ready to begin immediately. I wanted time to learn more about pedagogy and teaching before I began.

When I graduated with my Ph.D., my mom insisted that I let her purchase my academic regalia for me. She had been to several of college ceremonies and was always impressed by the complicated regalia worn by the faculty members. She knew that I wanted to teach and I would likely have an opportunity to wear my own regalia at least once a year as a professor. She likes to call the tam the "Christopher

Columbus Hat." I could tell how proud both of my parents were to see me complete my Ph.D. and I do think of them when I don my academic regalia for ceremonies each year.

Kim and her parents at UW-Madison, Ph.D. Graduation

Learning to Be a Professor, While Being a Postdoc (2002–2004)

Ray and I moved to New Hampshire so that I could start as Chris Bauer's postdoc. Ray now had 5 years of experience working for a Honda Dealership, and with his certifications, it was easy to find a job with another Honda dealership. We ended up renting an affordable house in a rural town about a half hour north of UNH. Ray and I commuted to work together to save money on gas, tolls, and wear and tear on our car. This half hour of time in the morning and the evening was a great time to catch up on things together. Meanwhile, working for Chris enabled me to meet even more amazing chemical educators from around the world. Chris supported my attendance at the 2002 Biennial Conference on Chemical Education (BCCE), which occurred at the start of my postdoc. I never knew I could learn so much about teaching college chemistry in such a short period of time. (I have attended every BCCE since then.) Additionally, Chris supported my participation in the UNH Preparing Future Faculty (PFF) program. This program was fabulous! I will be forever grateful for Chris' support in allowing me to participate while serving as his postdoc. I was able to take one or two courses a semester and learn by taking real courses about cognition and assessment. After completing my Ph.D. at UW-Madison, I was confident in my level of knowledge about chemistry itself. Now armed with this information about education and the scholarship of teaching I felt empowered and ready enough to apply for faculty positions.

Becoming a Chemistry Professor, the Academic Job Search

I applied for positions at predominantly undergraduate institutions (PUIs). I really liked to teach, and I liked doing some research, but I wanted to be at a place where the bulk of my time would be spent teaching and mentoring undergraduate students. I had several on-campus interviews with two in the same week at two different institutions in the greater Pittsburgh area. I liked the Pittsburgh area and California University of Pennsylvania was my top choice. Cal U was a state school, but the department was very interested in taking their General Chemistry program in a new direction. I felt my experience at UNH gave me the knowledge and desire to make positive changes to a program. Shortly after the interviews, the other Pittsburgh area school made me the job offer.

It was late in the week and I told Ray right away that we needed to visit the Pittsburgh area that weekend to see if he liked it, and would be happy living there, potentially for the rest of my academic career (at either school). When we moved to Wisconsin and New Hampshire we knew that they were both temporary moves and he trusted me to make the decision. I knew that now the stakes were much higher, and that he should be able to see the area and help me make the decision, as the impact could be more permanent. Ray seemed surprised at the urgency and that we needed to visit immediately. It was Valentine's Day weekend, and he had wanted to take me out to dinner to our favorite restaurant on Saturday night. We went to dinner on Friday, February 13th instead. Ray proposed to me that night and we were now officially engaged. I guess that while I was busy interviewing at six different schools for the past few weeks, he was busy ring shopping. We visited Pittsburgh and agreed that we would be happy to live in southwestern Pennsylvania. It was just a matter of fairly managing my job offers until I found out if Cal U was going to make me a competitive offer.

Cal U did make me an offer. It was a better offer than the others I had received and it was the place I most wanted to go. The advice of both of my mentors, both Art and Chris, suggested that it was important for me to negotiate. I had not formally negotiated any other job offer in my life, but I knew it was very important for me to do so. When I asked the Dean for a higher "step" on the unionized pay scale, he told me that the department was also hiring another faculty member, who had already accepted her initial offer. He asked if I still wanted him to find out if he could secure the higher rung on the ladder of the pay scale, and I said yes. He came back and agreed to the "step" that I had requested and told me that the administration was going to grant the other female faculty member the same higher step. So, I effectively negotiated for both of us! In 2004, my colleague, Dr. Yelda Hangun-Balkir and I were the first female chemists to be hired as Assistant Professors on the tenure track in the department. The first female physicist and Assistant Professor on the tenure track was hired by the department two years prior.

Pre-tenure, Pre-mother, Professor Years (2004–2008)

Ray started working at a Honda dealership when we first moved to PA. Several months later Ray left to pursue his own franchise as a Snap-on Tools dealer. These dealers travel local routes selling tools directly to auto-mechanics at their place of business. The first year of my position, I taught as Dr. Rickert. It was fun to have the same "name" as my dad, the Optometrist, Dr. Rickert. At Hartwick, many of our faculty members, like Bill and Susan, had us call them by their first names. I felt that I needed to gain more authority by having the students refer to me as professor or Dr. After my first year teaching, we got married, back in upstate New York surrounded by our families and friends from Hartwick, Wisconsin, and New Hampshire. I debated the pros and cons and in the end decided to change my last name when I got married. The week before I got married, I was invited to participate in the PASSHE Women's Consortium, Faculty Leadership Institute. I was so busy preparing for the wedding, and I naively thought, "These other women aren't chemists, what will we talk about for a whole week?" I didn't realize until that time how much we would all have in common as female professors. The workshop used the text, "Arming Athena: Career Strategies for Women In Academe" and one of the editors, Joan Chrisler, presented during the workshop. This experience was important for my development as a female faculty member. I not only learned valuable leadership skills, but I met an amazing network of supportive colleagues throughout the state system. My colleague Yelda left the department at the end of our third year. Her husband had career opportunities in another state so they relocated as a family. During her years at Cal U she was a valuable role model for me as she had arrived already a mother of a 3year-old son. I was able to watch her experience being both a professor and mother before I ever became a parent myself.

Kim's first day of teaching class. (Why do I look so tired already?)

The next couple of years, Ray and I both worked very long hours. I was updating the General Chemistry program at Cal U and Ray was running his own business. After a couple of years with Snap-on, Ray decided he wanted to return to fixing cars. While he was enthusiastic and highly knowledgeable about the product (tools), he wasn't enamored with the sales aspect of the job. He could identify with the clients so much and it was difficult to see some of them struggle financially. Ray returned to automotive work himself, working at a local independent repair shop.

In the original advertisement for the job that I responded to at Cal U, it described the idea of a chemistry "Studio." My Dean had immediately sent me on a trip to California Polytechnic Institute to visit their chemistry "Studio" facility developed and used by Tina Bailey. I began my research on the approach and worked with the architects during my first year or two at Cal U. Then the chemistry "Studio" project seemed to slow down as the administration searched for sources of funding to carry out the project. Even though the project was slowing down, my life as a professor was plenty busy. I was teaching a full course load of 12 credits per semester, and our department was hiring new faculty members and making curriculum changes to the entire chemistry major. Ultimately, patience paid off and the chemistry "Studio" facility renovations took place over the entire course of the 2007–2008 school year.

Birthing a Baby, and the Birth of a Chemistry "Studio" (2008–2009)

During the Summer of 2007, Ray and I agreed it was time to start a family. Being one year prior to tenure it seemed like it might be the "right time" if there ever was such a thing. I knew that I was getting older and that my job situation was very stable. I liked being at Cal U. Thankfully, I became pregnant very quickly and my due date was at the end of the academic year, April 2008. So, while the facility I had worked on for several years was being built, I met weekly with the team of folks doing the renovations. I always thought that we seemed an unlikely grouping of electricians, plumbers, carpenters, and a pregnant lady looking at blueprints and talking about circuits and HVAC. I always treated all of these guys with respect and I think I demonstrated enough mechanical and physical knowledge that the guys always treated me with respect.

I really enjoyed being pregnant. In some ways it was the pay-off for the inconvenience of menstruation for years. This was what my female body was designed to do. I was thankfully not experiencing morning sickness. I did worry whether my students or colleagues would hear me in the bathroom if these symptoms started. Thankfully, I was never sick. I also enjoyed having "company" with me everywhere I went. I didn't necessarily talk to my baby aloud, but certainly when I could start to feel him moving and he could respond to outside stimuli, it was a phenomenal feeling. Since I had become pregnant during the summer and he was due in April, I felt he should be born knowing both General Chemistry I and II and Inorganic chemistry because he had heard all of my lectures during the fall and spring semester. Walking around town I certainly remember the feeling of "we" instead of just "me."

Cal U had no formal maternity leave program in place, so when I explained to Human Resources (HR) that I was pregnant, they explained that I could take my earned days of sick leave or if I ran out of sick leave I could utilize the Family Medical Leave Act (FMLA) to have unpaid time off from work. HR made no suggestions to me, about what to do with my classes that were scheduled to end a couple of weeks after my due date. I am thankful to have great colleagues who offered to cover my classes for the last 2 weeks of the semester. This was uncharted territory for my department as no female faculty member had ever given birth during their employment. My goal was to work as long as I could until he arrived, so that I could minimize the amount of sick time I needed to use and limit the number of days of class that my colleagues needed to cover. I also didn't really want to sit home waiting for the baby to arrive. Even though I was exhausted, it seemed like it would be boring or stressful just sitting home and waiting for the baby. The uncertainty of when he would arrive was so foreign to me. I was so used to everything being programmed into my schedule and being able to carefully time and control when things took place. Giving birth is such a major event and yet I could not dictate when it would occur. What if my waters broke while I was teaching in class? In the lab? In the hallway? In my office? Like all first-time mothers, I was also nervous about the physical process of giving birth itself. Just three days before my due date my doctor induced my labor. I was showing symptoms of preeclampsia and they wanted me to deliver as soon as possible. So, my waters didn't break, I didn't go into labor, or I didn't have contractions at work. On one Friday I was on campus teaching classes, and the following Monday I was at the hospital being induced.

My son Max was born in April of 2008. I was just fifteen days shy of my thirty-third birthday. It was the end of my fourth year as a faculty member at Cal U. My colleagues Ali Sezer, Gregg Gould, and Matt Price only ended up covering the last two weeks of classes for me. During final week I wrote my own exams and a graduate assistant even dropped papers off at the house for me to grade. I am still grateful for the help of my colleagues. I didn't think it would be fair for me to submit for paid-sick days when I was really carrying out my final exam duties, with the exception the actual exam proctoring. Somehow I managed to get the doctors to write a note permitting me to return to work, even though they typically won't see you or verify this until six weeks postpartum.

Kim at home with Max (5 days old) working at the computer.

During the summer, I did not teach any courses, as I had opted to in the past. I was so happy that my position as a faculty member allowed me to stay home for a few months with my new baby. I would come in to check on the facility progress with Max in his stroller. The guys working on the room were still installing some of the finishing touches, and I think they enjoyed meeting the little baby I had been bringing in utero to the meetings we had had all year. I had never noticed parental accessibility issues until I brought an infant into a science building. We had a very large stroller, which Max's car seat snapped into, and it was too large to haul up the stairs to the second floor where my office was (we have no elevator). There is no changing station in the bathroom in my building, so I learned to manage. I would park the stroller downstairs. I would leave a towel, a few diapers, and wipes in my office, as well as a spare pillow in case I needed to nurse him while we were in the building.

Kim and Max (four months old) in the new "Studio Classroom"

That summer, I was scheduled to present my work on the chemistry "Studio" at the 2008 BCCE at Indiana University. Max was only 3 months old and I was not ready to leave him behind. He had been nursing exclusively, and I hadn't stored enough milk to be gone that long. So, my mom and my infant son came as my companions to the conference. We drove out with our pack-n-play and stayed at the hotel right on campus where many of the sessions were held It was a great way to stay professionally active and to show my mom what a chemistry conference was like as well.

Kim and Max at a 2008 BCCE Social Event

During the fall of 2008, Max started day care, as I returned full-time to work. At the start of the semester I missed his company when I returned to teaching and I no longer had my companion in utero. I wanted to continue to breastfeed Max as long as I could, so I became very familiar with pumping breast milk. I had a private office so this wasn't too difficult, but I would hang a "Do Not Disturb" sign on my door so that the custodian wouldn't come in to vacuum or empty my trash while I was pumping. Despite the sign, he did come in once, when I was thankfully turned away from the door, but in retrospect I probably should have explained the scenario to him, so he would take more care to avoid interrupting me. I didn't have a lot of time to dwell on returning to work with Max in day care, because at long last my chemistry "studio" had been completed. I was teaching for the very first time with a different pedagogy and my time at work was spent developing new classroom activities and experiments to be done in the new classroom. I did sometimes remark to myself that having a baby only took 9 months, while the design and renovation of the studio chemistry facility had taken several years. I was certainly proud of both of these accomplishments. My tenure and promotion dossier were due in November of that year. I had accomplished a lot before my son was born and was continuing to do more. I found it profoundly difficult to shift my time from "being productive" doing lots of teaching, research, and service, to spending time assembling the evidence to prove that I was "being productive." I really had wanted to breast-feed Max for a full year, but my milk supply was slowly diminishing. Most days I didn't have enough time in my schedule to pump more than once, and I was likely not staying hydrated myself since I was so busy. I was disappointed when I was only able to nurse him for about 7 months. We switched him over to formula until his first birthday.

In the summer of 2009, after my dossier was examined at all levels, I was officially awarded tenure. In 2009, I was also the first female chemist to be promoted to the rank of Associate Professor at Cal U. In the late summer 2009, the owner of the repair shop where Ray was working reported that he was going to downsize and work on closing the garage so that he could retire. We considered our options and Ray decided that he would stay home with Max.

Dr. Mom and Stay-at-Home Dad (2009–Current)

For the next two years, Max was home with Ray during the entire academic year. I cannot express how wonderful this was for me! The first year while Max was in day care, there was time involved in packing his bag (extra clothes, milk bottles, later food), dropping him off, and picking him up every weekday. I was fortunate to have a day care center located right on campus, but this daily routine amounted to at least one hour of time per day. With Max home with Ray, I could just wake up and sneak out early while they were still sleeping. I am so grateful that Ray was willing to try this out and that he also enjoyed it once he tried it. We still are all very happy with this arrangement.

I don't want to make it sound like utopia. The job of a chemistry professor is still quite demanding. I still spent loads of hours developing curricular materials, teaching, grading papers, advising the Chemistry Club, applying for grants, presenting at conferences, attending workshops and seminars, and serving on many university and departmental committees. It is a time-consuming job, but it is an interesting job. No two days are identical. The personalities and talents of each student are unique. The opportunities for research and collaborations evolve. The goal of continuous improvement of the courses, degree program, department, and university is always there. Having a family does not make the job more difficult or easier for me. I look forward to coming home to spend time with my family. It is difficult, sometimes, when I think about my work when I am home with my family, or when I think about my family (like a sick child) when I am at work. I have two different parts of my life that I find both challenging and extremely fulfilling.

I gave a presentation on the chemistry "studio" at the 2010 BCCE in Denton, TX. This time Ray and Max came with me.

Kim, Ray, and Max at the 2010 BCCE, Rodeo Event

Having Tenure and Having a Second Child (2010–Current)

When we first talked about having children we really liked the idea of having two. We each grew up with siblings and we wanted Max to have a sibling as well. While Ray and I have siblings that are each 4 years younger than ourselves, we didn't really think we should wait that long to have our second child. However, having a second child was more of a challenge than we had anticipated. My physicians assured me that I wasn't too old and that I needed to be patient. We were overjoyed when I did become pregnant and this time my due date was again in April. Being pregnant at the age of 35 meant that I was offered additional screenings during the early stages of my pregnancy. I was considered to be of "advanced maternal age." I did the same things I did the first time around for Max's pregnancy and I didn't opt for anything more invasive. I was so thankful that I was pregnant and I didn't want to take any additional risks for the purpose of screening. Thankfully, my pregnancy proceeded without major complications and my son, Samuel, was born 4 days before his older brother Max's third birthday.

In networking with other female faculty members on campus, after Max was born, I found out that other female faculty members on campus were able to get their colleagues officially paid for assuming responsibility for their courses when they had left for childbirth. Since I was the first female faculty member in the department to give birth, we didn't know on the department level how to go about making these arrangements when Max was born. So, while I was pregnant with

Sam, I made it a point to follow my campus colleagues' recommended course of action and get an official arrangement made so that my colleagues, Ali Sezer and Min Li, could be rightly paid for the courses they took over. I could then feel a lot less guilty about them coming in to take over my class. The arrangement was rather straightforward. Also, I wasn't sure that I wanted to work every last day until my due date the second time around. I already had Max at home and I was exhausted from not only work, but also from being mom to a toddler. I decided that this time around I would choose my last day of work, instead of letting my body and baby decide. I chose April 1st as my last day of work since I had an April 7th due date. I'm very glad that I did it this way. The uncertainty of when my colleagues would assume my duties was gone and knowing that they were going to be paid for their time made this much easier. The experience with my work arrangements was much less stressful this time. I didn't worry at all about writing or grading my final exams. The circumstance surrounding my son, Sam's birth, was somewhat similar. A few days after my due date, my doctor decided to induce my labor. With the advanced planning done at work I was much more able to enjoy those first 3 weeks of my son's life.

Chairperson Mom (2011–2014)

Three weeks after Sam was born, I took office as the Department Chairperson of the Department of Chemistry & Physics at California University of Pennsylvania. I was the first ever female Chairperson of the Department of Chemistry & Physics. The election for department chair occurred while I was pregnant and some of my colleagues had previously indicated that they thought I would do a good job in that position. When hearing that I was considering running for the position, another person on campus asked me if I would be able to handle the role with small children at home. I really do think that this was meant in the best possible way and I think that this person was genuinely concerned about the potential toll that this demanding position would take on my family life. However, if I were a male professor, whose wife was expecting a baby to be born around the same time, I'm not sure anyone would have asked me that question. I didn't really know how I felt about this. I wasn't sure if I should be disappointed that I was asked this question. (Hadn't I clearly demonstrated I could cope with challenges through my competence and commitment to professional responsibilities?) or if I should be sad for the professors who are fathers who might be in a similar situation and don't get asked this type of a question.

I did opt to run for the position, and I was elected. During the semester before Sam was born the outgoing chair helped train me on various items since I knew I would be out during the last few weeks of his term.

Kim and Sam at our Departmental Picnic the day she became Chair

When I did take office (with a 3-week-old infant) the department secretary had her office right outside mine. Our former custodian had retired, but our secretary could also turn away any would-be interrupters during my pumping times. Thanks to a quarter-time load reduction associated with being Chair, I was able to pump twice during the day, and I was able to nurse Sam until he was about 9 months old. I had a bigger desk, so I could just leave all of my pumping equipment and cleaning supplies there and I didn't drag the stuff back and forth from work to home.

I had the pleasure of attending the fall ACS meeting when my son Sam was only 4 months old. I had been appointed as an associate member of the ACS Women's Chemist Committee (WCC) and I had missed the spring meeting because it was so close to my due date. I didn't want to miss the fall meeting since I was eager to start working on the committee. Ray didn't want to travel with the full entourage to Denver, so I was diligent about pumping and freezing milk over the summer so there would be enough for a long weekend with me away. I studied the airline regulations regarding breast milk and pumps and I decided that I would "pump and dump." I would still pump so that I could keep up my milk production, but I wasn't going to try to keep the pumped milk food safe until I returned to Pennsylvania. I carried on my pump since I thought I might need to use it between the two legs of my flights. I didn't get any questions about it from the security. I did have it out and ready for inspection if necessary. There was a delay associated with the first leg of my flight and I didn't have enough time on the layover to pump so I actually ended up pumping in the bathroom during my second flight. I don't recommend it, but in the end it did help me to keep up my milk supply. During the conference itself, I brought the pump to the Convention center since my hotel was too far to walk to during the break. I remember leaving a bathroom stall with pump parts in hand to wash in the sink. I remember seeing Madeleine Jacobs, ACS CEO, in the bathroom. My initial feeling was of slight embarrassment. I guess I am kind of a modest person, but I knew I should instead feel proud. I doubt Madeleine even noticed me or the pump parts.

Kim and Sam (6 months old) October 2011

During the same fall semester I was headed to a conference that was still in Pennsylvania but the northeastern corner, about a five hour drive. I had packed my clothing bag, my toiletry bag, my backpack, and my breast pump bag and I had even remembered to download an audiobook, to listen to as I drove. An hour into my drive, my husband called and told me I had left my clothing bag at home on the bed. I had made multiple trips to the car and had apparently failed to make the last trip needed. How could I forget my clothes??? I had my laptop and my work papers, I had my toiletries, and I had my breast pump supplies in that bag, but I didn't have a stitch of clothing except what I was wearing (which was fairly casual since knew I would be sitting in the car so many hours). Had I left behind my pump, I probably would have turned around and driven back to get it, even though I would have lost two hours time doing so. Since it was only clothes, I went shopping. I stopped at a women's clothing retailer I saw in a strip mall. It was already near six o'clock in the evening and I knew that if I waited until I got to a bigger city or closer to my destination, stores would be closed. When the store attendant asked if there was something in particular I was looking for, I was speechless. Luckily for me there were sales and this retail chain wasn't that expensive to begin with. I got tops, pants, pajamas, and undergarments, even a pair of shoes. While this was fun and something I will never forget, I will also never again forget to bring my suitcase on a trip! But, I still pumped my milk and stored it in the hotel fridge and was able to bring it back when I returned.

In the Summer of 2012, I resumed teaching summer class. I was the department chairperson and that required that I supervise the staff, write reports, and process orders so I may as well have been teaching (which is much more fun than most administrative duties).

My three-year term as the department chairperson ended in April 2014. I did not opt to run for a subsequent term. I have a colleague who was willing to serve and was been elected to the position. I am glad that I did serve as the department chairperson because it is a position you don't understand well until you experience it firsthand. I learned a lot about how the university works and I've gotten to know the administrators on campus very well. It is a very, very, time-consuming job, if you accept the full responsibility involved. It involves solving a lot of problems or developing new policies and my colleagues indicate that I did a favorable job. I decided it was just not right for me. I will entertain doing it again some time in the future.

As Max and Sam get older, we continue to navigate changes to our family routine. When Max was four he started a preschool program on my campus. The preschool program was operated by the Department of Communication Disorders and focused on Language Skills. Two faculty members and a cadre of undergraduate and graduate students in speech pathology administrated the program. Since the program was actually associated with a college course, it started promptly at 9 am—no sooner. When the department schedules were being made up, I didn't realize it would be problematic for me to drop him off and get to my own 9 am class (in another building) on time. I take my punctuality quite seriously and hate to be even a minute late for any of my classes. After speaking with the faculty members who ran the program I was able to drop Max off a little bit early, but they didn't want the other parents (who weren't on their way to class) to see this. Had I had the knowledge earlier, I could have scheduled my own classes for a different time slot. Sure, Ray could have brought him in to the program, but he would have had to wake and dress Sam and buckle him into the car seat, drive to campus, unbuckle both kids, and walk them in. Then he would have needed to repeat the process three hours later when the program was over at noon. Instead we opted to make the best of it with me dropping Max off and Ray picking him up.

I think it would be safe to say of both parenting and professorship that just when you have things figured out, they will change. There will be more hurdles to overcome, and life will get more hectic when both kids have school, homework, swimming lessons, and soccer practice, but working together as a couple, Ray and I have always found a way to make things work, and I'm sure we'll continue to do so. I have to give a lot of credit here to my amazing husband, Ray. When we met in high school, neither of us had any idea our life would evolve this way. He is truly a wonderful father and spouse who I'm happy to be spending my life with and he has always been immensely supportive of my career.

The Woznack Family, Indoor Waterpark Vacation, Spring 2013

Lessons Learned and Advice

- No one is perfect. All human beings have twenty four hours in a day. It is a challenge to any mother or parent who works full-time. Women should be honest about the multiple roles they play in many people's lives and the time those roles require. We cannot keep up the illusion of "effortless perfection," when all working moms orchestrate a delicate ballet of responsibilities every day.
- Even though we have finite hours in the day, that doesn't need to impact the quality of the time we spend at home or at work. My advice would be that as much as possible, when you are interacting with your children you should be PRESENT. As a faculty member, there are many hours where we cannot physically be at home, but when you are home, try to make the most of every moment by focusing on your children. I have ditched using my smart phone at home as much as possible because it kept distracting me from being "home" when I was home. Likewise, I don't stay plugged into the phone while I am at work. Ray and the Elementary school have my office number and will call on the landline when there is an emergency or something important.
- Ask for support from others when you need support. This is important both at home and at work. If you need assistance with a work-related project, it is usually acceptable to ask a colleague to collaborate with you so you can split up the work. Likewise, an understanding spouse will understand if you might be delayed from your household responsibilities.
- You are your own harshest critic. Give yourself space to be human. Have realistic expectations for how much work you can complete at what level of quality in a certain period of time.

- Compliment other people (especially women) who impress you with their professional and personal accomplishments. The people you are impressed by are likely their own harshest critic and sometimes it takes an outsider to point out the obvious accomplishments. It just might make someone's day to hear that you noticed and are impressed by what they have achieved or a decision they've made.
- Surround yourself with a supportive network of both colleagues and friends. It is important to trust your "gut" or instincts when interviewing for a position. An academic position can be a work equivalent of "home" for a very long time. Try to select a supportive environment. Also find support outside of your department from other like-minded colleagues. Find support among your nonacademic and non-chemist friends.
- Embrace your identity (and perhaps inner Geekdom). I will be honest that when I meet people outside of the academic environment, I don't always start out with, "I'm a Chemistry Professor" because that sometimes is a conversation killer. People will possibly report back that they hated chemistry or they did poorly and where do you go from there? I know female faculty members in other disciplines (such as psychology) have an equally difficult introduction. Come up with a response you can use that is neither derogatory nor self-deprecating. When helping out at my son's kindergarten holiday party this year, the teacher asked him to introduce his mom to the class. He said, "This is my mom and she is a chemistry teacher and her students just graduated." I was proud that he knew what I do for a living and was happy to tell his classmates. I was surprised he added the comment about graduation, but I had indeed gone in to the ceremony on a Saturday morning (in my regalia) and Max had noted my absence.
- Be proactive about your course schedule. Even if you are not comfortable telling your colleagues that you are beginning to plan a family, you may want to think about your class schedule. If you are in a small department like mine, colleagues won't be able to cover your courses if they are all meeting during the same hour of the day. If your child is about to start preschool or kindergarten, find out when those programs start or end so that you can minimize conflicts with your course schedule.
- If you do have the pleasure of being a mother and a chemistry professor, be a role model for your female students. I have snacks and crayons and coloring books in my desk drawers so that in any type of emergency or when needed my sons can come to work with me and spend time being creative. My sons love wandering the halls of the building and climbing on the benches and using the vending machines when they visit. I can't wait until they are a little bit older and I can actually bring them into the labs to do more! When our campus marketing department asks if I am willing to participate in a photo shoot or video, I try to almost always say yes. I like the idea that they want a female chemist to be visible in their marketing campaign.

Main Steps in Kim's Career

Education
Ph.D., Inorganic Chemistry, August, 2002
University of Wisconsin-Madison, Madison, WI
Thesis Title: "Surface chemistry and electronic properties of GaN studied via synchrotron-radiation-based photoemission spectroscopy"
Advisors: Arthur B. Ellis (Department of Chemistry) and Thomas F. Kuech (Department of Chemical Engineering)
B.S., Chemistry, May, 1997.
Hartwick College, Oneonta, NY
summa cum laude, College Honors, and Departmental Distinction

Academic Positions

California University of Pennsylvania (2004–present)
  Department of Chemistry & Physics, 250 University Ave. California, PA 15419

- Department Chairperson, May 2011–April 2014.
- Associate Professor, Fall 2009–Present
- Assistant Professor, Fall 2004–Spring 2009.

  Postdoctoral Researcher, August 2002–August 2004.
  University of New Hampshire, Durham, NH 03824.
  Project: Integrating the Chemistry Systemic Initiatives
  Advisor: Christopher F. Bauer, Department of Chemistry

Acknowledgments I could create a long list of people who have supported me at various points throughout my career, but I want to especially thank my biggest supporters. I have to of course thank both of my parents for being such amazing role models. They taught me to strive for excellence and to have a strong work ethic. They taught me how to be professional and they have demonstrated how strong a partnership a marriage can be. They truly always encouraged me to succeed. I also have to thank my truly devoted husband, Ray. He is patient and kind and understanding, of my unique position. He has always supported me in my career, at every turn. I cannot express in words how grateful I am to have found such a wonderful companion in life. I also want to thank by children Max and Sam for their unconditional love. It is a wonderful feeling to return home from work to see their smiling faces and hear them say, "Mom's home!", which they say while running up to give me a hug. I also need to thank Ray's family for all they have done and continue to do to support our family.

Remarkable, Delightful, Awesome: It Will Change Your Life, Not Overnight but Over Time

Sherry J. Yennello

She's remarkable! I said that when she was born and many times since. Stephanie was born with ten perfect little fingers and ten perfect little toes. She came out crying and all I wanted to do was soothe her and take away whatever was causing her to cry, but the nurse told me it was good for her lungs. Thus began a remarkable journey that has changed my life.

Growing Up

Let me step back to my own childhood. I had magnificent parents and two older sisters. My mother earned a college degree, but, like many of her generation, she stopped working when my oldest sister came along. My father dropped out of school in the 6th grade to join the Merchant Marines. When my sisters and I were still quite young, he went to night school to get his GED. He didn't go back to school for career advancement; he did so to send a message to us about the

S.J. Yennello (✉)
Cyclotron Institute, Texas A&M University, College Station, TX 77843-3366, USA
e-mail: yennello@comp.tamu.edu

importance of an education. There was never a question of whether any of his daughters would go to college, just a question of where. When my time came, I went to Rensselaer Polytechnic Institute. I got dual degrees in chemistry and physics. My father always enjoyed parents' weekend, because although he had little formal education, he appreciated all the engineering demos that were always on display. My father was—and is—a very smart man.

While going through college I didn't really have a plan for my future. In fact it wasn't until the summer after I graduated, while working at a nuclear power plant and getting ready to go to Indiana University for graduate school, that I realized I wasn't going to live the life of my mother. I was sitting by the locks in Fulton, NY, recalling a conversation I had the previous year with a friend from high school. We were talking about our futures. Even though I had worked hard through college, I said I would give it all up for the right guy—after all, that is what my mother did. My friend told me "for the right guy, you won't have to give it all up."

The Right Partner

After graduate school and a postdoc at the National Superconducting Cyclotron Laboratory at Michigan State University, I accepted a position as an Assistant Professor of Chemistry at Texas A&M University in College Station, Texas. My career as an experimental nuclear chemist was starting, and I was very excited. And...I met the "right guy." Larry was also a young faculty member in the chemistry department. He was—and is—a guy for whom I would give it all up, but for whom I don't have to. He has been tremendously supportive of my career, sometimes at the expense of his own. He has been there at every step; he went to every doctor's appointment when we were pregnant.

The Decision

In my mind, being an assistant professor and an experimental nuclear chemist presented a huge obstacle to becoming a parent. How would I balance my career and a family? How would I avoid the radiation area where I do my experiments for a nine-month pregnancy? I had great career role models in two senior nuclear scientists Ani and Jolie, but neither of them had kids. Two amazing women helped me see that being a parent and being a scholar are not mutually exclusive. Shirley Jackson, then president of my alma mater, visited my campus as part of a Women in Discovery symposium. While escorting her from one meeting to another she explained to me that I could do an experiment without being in the radiation area as long as I had a group of people with whom to work. Additionally, Geri Richmond visited campus and also provided great encouragement about being a mother and an academic. She assured me it could be done, as long as I was willing to ask for help.

These two incredible women probably don't even know how much they impacted my decision to become a mother. Now I just had to figure out the best time.

Timing

We had been married for about five years and my tenure had been granted when our department made the offer to let faculty double teach—something previously much frowned upon—under certain special conditions. They were trying to get more of the tenured and tenure-track faculty into the larger service courses, because it looks better to uninformed people who don't appreciate the talent that exists in the non-tenure line faculty. So they offered to let any tenured or tenure-track faculty member have a semester off in return for teaching a larger service course in addition to a regular teaching assignment. They assumed that this would enable the faculty member to have an uninterrupted semester devoted to research. When I took the deal, the associate head assumed that this would just mean we wouldn't schedule nuclear chemistry (a small upper division course I taught once a year) the upcoming fall. I said no and that I wanted the following spring semester off. He queried "Why?" and I told him because that was what I wanted. How could I tell him I wanted to have a kid when I hadn't even discussed it with my husband yet?

So I think I shocked my husband when, over dinner at a favorite local restaurant, I asked him if we wanted to have a child. The timing was perfect; I had tenure and I had arranged a semester off. Fortunately for all of us, he agreed and we made a decision that would change our lives. We knew I would have the spring semester off, so we calculated backwards and decided when we should start trying to get pregnant.

Pregnancy

Biology was good to us and our child was scheduled to arrive just after finals in December. When Larry and I were in San Francisco for an ACS meeting I had a suspicion that we might be on our way to becoming parents, but it was too early to tell for sure. Nonetheless, despite a very nice dinner at a very nice restaurant, I passed on the wine—just in case. I didn't want to do anything that wasn't in the best interest of my future child. Additionally I gave up Twizzlers and all other junk food and ate fruit for snacks for the next nine months. Nothing but the best for my future child.

When we first knew I was pregnant, Larry was ready to tell the world. I wasn't ready to let the people in our department know, however. There were very few women among the tenured or tenure-track faculty, and none with kids. So we told our families, but otherwise kept our news to ourselves for a number of months. I was being creative about how to avoid the radiation area, but I finally had to break my silence and tell the cyclotron laboratory director because a congressman was visiting campus as part of a science and public policy program and he wanted a tour. I had been sitting near him at lunch and did what I almost always do with people I meet, which is tell them they should come see the cyclotron. I knew his day was crammed and he wouldn't have time for the tour, so I didn't think anything about my offer. But by the time I walked back across campus there was a phone message that his schedule had been rearranged and he was headed over for a tour. I panicked. How could I go into a radiation area to give a tour? How could I get someone else to give the tour? How could I tell the Congressman that I couldn't give him the tour? So I settled on asking the cyclotron director to give him the tour; his status made it seem OK that he was to give the tour I had offered. But the price was I had to break my silence. The director promptly told his wife, and I learned that even solemnly sworn secrets come with a spousal exception. Fortunately it was still a number of months before my pregnancy became known in the department.

Although both Larry and I believe that information is good, sometimes too much information isn't a good thing. Since we were older parents we opted to have an amniocentesis. We consulted with a neonatal specialist, who said you really shouldn't do the amnio until 16 weeks, but she did a sonogram to see if there was any reason for concern. The nuchal translucency she measured put us at a slightly higher risk of complications. I went insane. This rational scientist who spends a lot of time looking for small signals and understanding probabilities couldn't be rational. This was my future child. I needed to be certain that she was perfect. Larry convinced the doctor to do the amnio a week or so early, and we were very happy when the news arrived that we had a perfectly normal daughter on the way.

Having told my research group that I was pregnant, I would go in the cave to help set up the experiment, but once we had beam I would confine my involvement to the counting room. My next challenge was that I was teaching the nuclear chemistry course in the fall (part of the condition for getting the spring off). Part of the course was a laboratory where we used radioactive sources. The exposure would be minimal, but I didn't want to risk the health of my child in any way. It seemed reasonable that I could teach the lab without actually being in there when the sources were being used. In order to do this, I would need a little extra help from a graduate student. I asked the department head for 1/3 of an extra teaching assistant so I could replace my physical presence during the lab with the graduate student who had taught the course with me the previous fall. Fortunately, the department head agreed.

Graduate student assistance in my nuclear chemistry lab was the only accommodation for my pregnancy that I asked for from the department—I had already "bought" my semester of teaching relief. However, I was contacted by the HR liaison in the department to inform me about my right to take time off under the Family and Medical Leave Act (FMLA). (She hadn't felt the need to inform any of the male faculty who had recently had children.) She instructed me to get a doctor's note so I could use six weeks of sick leave until I had to invoke FMLA. I explained that my child wasn't due to arrive until after finals and I had arranged to have teaching relief for the following semester so while I appreciated her advice I didn't think I needed to invoke any of it. She took a calendar and informed me that faculty and staff were supposed to work until the 22nd of December. I assured her that there would not be a day in which she could find 50 % of my faculty colleagues in their offices that I would not be accounted for. When I asked if she had done this briefing with my male colleagues who recently had kids she told me she was just treating me like "any other pregnant female in the department." The problem was that her mental model for a pregnant female was a staff person for whom she would hire a temporary replacement while they were on maternity leave. My problem was I wasn't sure where she thought she could find a temporary replacement to run my research group. She followed up our meeting with a memo to me repeating all her instructions about getting a doctor's note and invoking FMLA. After calming down (Larry was very helpful here) I wrote the department head an e-mail saying I hadn't asked for any leave, but I would let him know if at a later date I thought I needed to take leave. He must have asked her to step back because that was the last discussion I had with her about maternity leave.

While I was pregnant I was informed that I had been selected as the Sigma Xi National Young Investigator, but I would have to accept the award in person in Albuquerque and give a talk at the national meeting. The meeting was scheduled one month before I was due. I accepted and Larry made plans to travel with me to the meeting in case I needed any help. My health insurance plan prohibited travel outside of a 90-mile range in the last month of pregnancy, and my award talk was just outside that window. But about a month before the meeting my OB said she thought my daughter would arrive about a week earlier than the original due date. Fortunately I convinced her not to move my "official" due date so I could accept the award. Larry was fabulous and even found a suit for his very pregnant wife to give a talk in.

My Daughter

Stephanie did arrive "early" as predicted and was born with jaundice. Fortunately, my last lecture had been given and my exam had already been written and printed. I called my teaching assistant who agreed to do the review for the final and give the exam. She would later deliver the final exams to my house so I could grade them. Stephanie spent the first 5 days of her life in the Pediatric Intensive Care Unit under special lights to treat her jaundice. I took a week of sick leave. My parents, excited about their first (and still only) grandchild, flew in on the red-eye. They were amazing as they took care of Larry and I. We spent those first days driving back and forth to the hospital for feedings. Fortunately, Stephanie was able to come home before my parents had to leave.

The first weeks after Stephanie was born were eventful. Not only were we dealing with this new life, but my first Ph.D. student, Doug, was set to walk the stage at graduation. Stephanie stayed home with her dad, but she gave me the perfect reason to leave the ceremony shortly after I had presented Doug with the hood symbolizing his degree. The following week, Larry and I had a faculty meeting to attend and we took Stephanie with us. The associate head was caught off guard. He said he didn't know that I was coming back to work—as if having a child meant the end of my career. It didn't mean the end of my career, but it did mean we needed to get some help.

Childcare

At first Larry and I thought that we might be able to arrange our schedules such that we could alternate who was home to take care of Stephi—particularly since I had a semester of teaching relief. It quickly became clear that our plan wouldn't work, regardless of a newly upgraded computer at home. So we made a plan for how one might go about finding a solution to our dilemma. Not a list of possible solutions, but a list of steps we could take that might lead to a solution. One of the first steps was to ask some of our friends who had young kids how they dealt with the childcare issue. We were told about this wonderful nanny, Dorothy, who had previously been employed at a childcare center in town, but then went on to be a nanny in Houston. Now she was coming back to town so we got her contact info and arranged to meet her. Within 5 minutes we knew she would be perfect. Dorothy took care of Stephanie for the next eight months, which gave us just enough time to get our bearings before the next chapter of our lives started.

When Stephanie was 8 months old, we moved to Washington, DC, to work for the National Science Foundation (NSF) for 1 year. We spent 3 days driving to Arlington, Virginia. When we arrived, we put Stephi in group care for the first time. Bright Horizons, a childcare center, was on the first floor of the NSF building. Luckily a wonderful staff member at NSF, Denise, had connected me with the center, and we got on the waiting list months before we moved. Since the NSF had priority at the center, we were able to actually get Stephi a spot. We cried a lot those first times we dropped her off. The worst day was when I had to go back to College Station for an experiment the week after we moved. I flew back to Texas leaving my beautiful eight-month-old daughter in a world that had been turned upside down. The only familiarity for her was her dad, thus starting an incredible bonding between the two of them. She could only fall asleep curled up on his chest for the next year. Returning to Texas for that experiment is a decision that I would make differently if I could roll back the calendar because it took over a year for Stephi to forgive me for leaving so early on.

My sisters thought that one of the biggest changes we would have to go through with a baby was giving up going to nice restaurants, which was something Larry and I certainly enjoyed. However in October, when Stephi was ten months old, we ended up in Hawaii and had to figure out how to feed her in a restaurant. After this went well, and we were back in DC, we proceeded to take her to all the top restaurants in DC. Many a waiter marveled at what the baby would eat. Some years later she became quite picky about what she would eat, but just this year she has become a vegetarian and is a most adventurous eater once again.

As the year went on, and I took trips back to College Station to run more experiments, I started checking out childcare centers again so we would be set when we returned to Texas. This was when I realized I needed a lot longer than I thought to get into a childcare center—you were supposed to get on the waiting list before you were pregnant if you wanted a spot before the child was two. Fortunately, we were able to get Stephanie into the TAMU children's center.

Travel

At seven weeks old Stephi took her first airplane flight. Larry and I had a meeting to attend at Michigan State University. We had a friend there who arranged for someone to help look after her while we worked, and I was able to find an empty classroom in which I could pump. It was not much longer after that trip that I traveled alone to give a talk at Arkansas State University. I lacked the confidence to ask for time and space to pump, so I ended up sitting on the floor of a bathroom.

When Stephi was four months old there was an American Chemical Society meeting in San Diego, and both Larry and I were scheduled to go. We had decided I would fly to my parent's home in Oregon with Stephi and spend a few days so everyone could get acclimated with one another. Then I would fly down to California to meet Larry. Larry was very sad the night Stephi and I left for the airport. He was about to face his first night without his daughter since we brought her home from the hospital. When my parents dropped me off at the Portland airport a few days later, I was equally sad since I was leaving her, too. I took a picture of her and left in tears. As soon as I landed in California I sought out a 1-hour photo developer (we weren't yet in the digital age). We had that picture—if not our daughter—for the next few days. At the end of the meeting we both flew to Portland and couldn't have been happier to have her back in our arms. We made the decision then never to go to the same meeting without taking her with us. Now when we both go to the same conference, we creatively merge our schedules and we arrange to spend time with her, too. We just needed to coordinate our schedules. Communication and coordination are important every day. I don't agree to meetings before 8 or after 5 and never accept a travel invitation without discussing it with Larry. Twelve years later I'm still sad every time I head to the airport without her.

As Stephanie grew older it became more important that she knew ahead of time that one of us was going to be traveling. I would sing a refrain to her many times:

Mommy will always come back,
Mommy will always come back.
Mommy loves you very much and
Mommy will always come back.

I also left her notes to find while I was gone. Eventually, Stephi played a game where she would walk around the house pulling her small suitcase and say she was going on a business trip. She said she would call when she got to the hotel. The hotel was under the dining room table.

Professional travel is something that is always a balancing act. I travel less than I would without her and feel more guilty when I do travel. For her first birthday I was supposed to be at a meeting in Bologna. I don't know what I was thinking when I agreed to the trip and booked the plane tickets. But while I was there I realized what a mistake I was about to make, so I switched my plane tickets and came home early. In later years I was smart enough—and comfortable enough—to just say no to trips that conflicted with her birthday. I've also taken red-eye flights home from meetings on the west coast so that I could make it to her soccer game the next morning in Houston. I have scheduled flights to be after her games—in one case getting to the airport in just enough time to take a shower since I was on an overnight flight to Europe (and I needed to get rid of the sweat and sunscreen). My travel planning always involves several questions. Do I need to go? What is the last flight I can take and get there on time? What is the first flight I can take to get home after I have done what I needed to do? Sometimes I Skype into meetings rather than travel. Regardless I am always sad when I head to the airport. I love my daughter and don't want to miss a single day of her life.

Advice in a Nutshell...

- Find a supportive partner or other support network. My husband is wonderful.
- Know your environment, or more importantly choose one that is going to be supportive. That could mean picking your thesis advisor or postdoctoral mentor. It definitely means carefully picking a department that is going to help you reach your goals.
- Timing is everything. There may never be a perfect time unless you create it.
- When you get to choose who you work with, choose wisely.
- Communication is critical. Don't assume that your partner, your child, your students, or your colleagues know what you are thinking or planning.
- If you make a decision you can't live with, unmake it.
- Be creative about travel. Do you need to go? Can you Skype in rather than attend in person?

She Has Changed My Life

My daughter is delightful. She is 13 now. I have had the privilege of watching her grow and learn and watching her tackle new challenges and become the most amazing person I know. I often wonder how she got to be so awesome. I could not imagine life without Stephanie. I try every day to appreciate the gift that is my daughter.

Main Steps in Sherry's Career

Dr. Sherry J. Yennello, Regents Professor of Chemistry, Director of the Cyclotron Institute and holder of the Bright Chair in Nuclear Science at Texas A&M University, is an internationally renowned nuclear chemist. Sherry joined the Texas A&M faculty in 1993 after serving as a postdoctoral fellow at Michigan State University (1991–1992) and earning her Ph.D. from Indiana University in 1990. She was architect and cochair of an NSF-funded Gender Equity Conversation effort and served for many years as the chair of the College of Science Diversity Committee. Sherry is a fellow of the American Chemical Society (2011), the American Physical Society (2005), and the American Association for the Advancement of Science (2013). Her many awards include the ACS's Francis P. Garvan-John M. Olin Medal (2011), the Texas A&M Women's Faculty Network Outstanding Mentor Award (2010), the NSF Young Investigator Award (1994), and the General Electric Faculty for the Future Award (1993).

Part II
Safety in the Lab

Safety and Motherhood in the Chemistry Research Lab

Megan L. Grunert

As more and more women enter the workforce, issues regarding childbearing and caretaking have become more prominent. Challenges women face with the timing of pregnancy, finding affordable childcare, and meeting the multitude of demands inherent with raising children are starting to be addressed by many employers. In the university setting, structural and cultural factors pose barriers, yet also provide almost unheard-of flexibility. While some of these challenges are being addressed by employers, funding agencies, and communities, one area of particular concern is safety for women chemists. The hazards of working with chemicals during pregnancy and breast-feeding are unique to chemistry, and there is no easy fix. This chapter will review the safety concerns for women chemists, strategies women faculty have used to address safety challenges, current policies and strategies for dealing with safety challenges, and recommendations for departments and administrators.

M.L. Grunert (✉)
Department of Chemistry & The Mallinson Institute for Science Education, Western Michigan University, 1903 W. Michigan Avenue, Kalamazoo, MI 49008-5413, USA
e-mail: megan.grunert@wmich.edu

Safety Concerns for Mothers in Chemistry

Chemists are familiar with safety training, and highly publicized cases, including a fatal accident at UCLA, have recently led to an increased examination of safety practices and appropriate training. Often not addressed in safety training are issues for pregnant and breast-feeding women. Obviously, many hazards are directly associated with the chemicals used in the laboratory setting. Many organic solvents, heavy metals, and other hazardous compounds pose risks to developing fetuses and infants. Material Safety Data Sheets (MSDS) provide information regarding known hazards, but what about unknown hazards? MSDS do not make recommendations on protection or exposure levels, merely providing what type of hazard a chemical poses, if it's known.

One of the greatest challenges for women chemists who are pregnant or breast-feeding is the lack of standardized information. The Centers for Disease Control (CDC), the National Institutes of Health (NIH), the Environmental Protection Agency (EPA), and some university environmental health and safety offices have information about pregnancy hazards publicly available; however, this information does not include all chemicals used in research. These databases cover common hazardous chemicals, medications, and illegal substances. A comprehensive list is currently unavailable. The Agency for Toxic Substances and Disease Registry (http://www.atsdr.cdc.gov/toxfaqs/index.asp), maintained by the CDC, maintains a list of fact sheets addressing frequently asked questions about many chemicals. As with many references, it is not overly helpful with regard to pregnancy and breast-feeding. Consider the entry for acetone, where the last paragraph under the heading, "How can acetone affect my health?" says, "Health effects from long-term exposure are known mostly from animal studies. Kidney, liver, and nerve damage, increased birth defects, and lowered ability to reproduce (males only) occurred in animals exposed long-term." For a woman deciding whether it is safe to be around acetone while pregnant, this is so vague as to not be helpful. Similarly, the page for mercury includes the statement, "Exposure to high levels of metallic, inorganic, or organic mercury can permanently damage the brain, kidneys, and developing fetus." It is unclear what "exposure to high levels" means, although the outcomes are clearly severe. The National Institute for Occupational Safety and Health, part of the CDC, includes a statement in their document The Effects of Workplace Hazards on Female Reproductive Health, about exposure to hazardous chemicals during pregnancy, "Whether a woman or her baby is harmed depends on *how much* of the hazard they are exposed to, *when* they are exposed, *how long* they are exposed, and *how* they are exposed" (http://www.cdc.gov/niosh/docs/99-104/pdfs/99-104.pdf, p. 14). Again, it is a less than helpful warning for women chemists.

While working in a research laboratory, women have to look up possible hazards on the MSDS or through other available reference lists for any chemicals they are using. Either working with a safety officer or on their own, they then need to make a decision about whether they want to expose a developing fetus or breast-feeding infant to these chemicals. Often, these decisions are strained by a lack of information about what specifically poses risk, what levels are dangerous, how mobile chemicals are in relation to skin and cell barriers, and what personal protective equipment would help and how effective it is. This can effectively put women chemists in a position of choosing their career or the health and development of their child(ren).

Strategies Used by Women Faculty to Address Safety Challenges

The impact of motherhood for women chemists is different depending on the type of institution, the teaching load, research expectations, and service responsibilities. While it might seem that motherhood would pose a greater challenge to women faculty members at research-intensive institutions, there are significant challenges to women at primarily undergraduate institutions (PUIs) as well. At research institutions, research productivity can be somewhat maintained by postdoctoral associates and graduate students in the absence of the research advisor/primary investigator (PI). The National Science Foundation (NSF) has also instituted policies allowing for grant funding to be used to hire a lab technician or for the grant to stop for a year. At PUIs, faculty members may have to maintain an undergraduate research program. In the absence of graduate students or postdocs, it falls to the PI to train student researchers and work with them in the lab. Thus, a pregnancy can force PIs out of the lab, essentially halting research progress. This is a challenge after pregnancy as well, when the PI tries to restart her research program.

One PUI faculty member, Ellen, took five years off from undergraduate research during two pregnancies due to the use of a mercury bubbler. While her department adjusted her workload by removing the research component and adding instructional and service responsibilities, this did not address the stalled research progress or the difficulty in restarting her research program. When telling her story, she said,

> ...I'm back this year for the first time basically in...five years, because I had two kids. So I was out of the lab [for] more or less five years straight, [because of] pregnancies and then breast-feeding, and then trying to get the second one, and...so I'm finally back. I was a big believer and didn't go near [the lab] the whole time...I'm an organometallic chemist who uses a mercury bubbler! Academic institutions do not understand how to deal with female faculty members, especially in the sciences, who choose to have children. They're just confused with how to deal with that. ....There's no road map...the difficulties of dealing with chemicals and the campus was very, very confused about ...how to deal with that. That's probably been the biggest challenge. ...I picked up other teaching duties, I took responsibility of making sure that all the general chemistry laboratories were [updated], the lab manual, that became my baby. Writing the whole thing, proofing the whole thing, dealing with our stockroom support on making sure the labs get prepped, that became my job, but then I also picked up additional teaching load, and that I worked out with the dean [to compensate for not being in the research or teaching labs].

As she points out, there is no "road map" for science departments to address the concerns of women chemists who are pregnant and/or breast-feeding. Luckily, she was able to work out a compromise with her overall workload, but she went on to discuss how challenging it was to return to the undergraduate research lab.

Another PUI faculty member, Laura, discussed challenges associated with teaching organic laboratory courses while pregnant. At PUIs, faculty members are often the laboratory instructors, rather than graduate teaching assistants as is standard practice at research universities. She discussed feeling uncomfortable being in the teaching lab while breast-feeding as well as being limited in the research lab.

> I think the number one thing is that if you're an organic chemist, and, really I would say a wet chemist, I don't think it applies to all fields, but it's a challenge, child-bearing. As a chemist, you know, you're doing wet chemistry or using things sometimes that you don't wanna expose your kids to. If you teach labs, then that's nine months that you're technically not teaching organic lab. In terms of, breast-feeding your child, I was in the lab when I was doing that, and every day I felt sick [for] my child. Subsequently the other female faculty did not teach with their second children, in the lab, while they were breastfeeding. So really that pretty much gives you two academic years almost where you are not teaching the lab, and it will limit the time you spend actually in the research lab also. I think time-wise, certainly at an undergraduate institution you don't have to work as long [in terms of] hours, but in terms of the impact child-bearing [has] on the ability to do your job, [it] is entirely different.

Her concerns about how pregnancy and breast-feeding impact female professors' careers echo those stated earlier. She discussed feeling an obligation to uphold her teaching responsibilities, but later wished she had worked out an alternate solution.

While Laura's colleagues changed their post-pregnancy plans based on her experiences, Irene worried about how her decision to not take maternity leave would affect her colleagues. She had no regrets about not taking time off, as her department was understaffed at the time and she worried about overburdening her colleagues. Irene did not express the same safety concerns as Laura and Ellen, but she is a plant biochemist and believed the chemicals she worked with safe to be handled during pregnancy. She described her experience, saying,

> I think being one of the first faculty women, there's also pressure that you do it [and] do it right, so that you set a good precedent. I'm somewhat concerned about the precedent I've set with having kids since I didn't really have maternity leaves for two of the three, we did not hire a replacement for me, so that put pressure on my colleagues. And I'm now in the position [where] younger women faculty ask for advice. Am I giving good advice? I don't know. Well, I can tell you what I did, [but] I don't know if it's the best answer, or what I would do differently.

As seen previously, she expressed doubt about her decisions. Lack of clear policies, recommendations, and precedents makes navigating pregnancy and breast-feeding challenging for women chemists.

At research-intensive institutions, women expressed more concern about the timing of pregnancy with regard to tenure rather than safety concerns with chemicals. They commented on a lack of structure or support within the university regarding maternity leave and arranging teaching assignments, concerns regarding postponing pregnancy until after tenure, and having fewer children than they would have liked due to their careers.

Petra, a professor at a research-intensive university, discussed some of the challenges she specifically had while pregnant. She commented on the university structure at large and a general lack of support for mothers. She believed the hurdles for mothers in the university setting were part of the reason women left research institutions. She said,

> ...teaching assignments when I was pregnant, arranging maternity leave, there were just lots of discussions and issues that made these arrangements not ideal, so I think there is a lot of work to be done, [and] this was a little bit a cause of grief...there is so much to be done to make this job for a woman more human and more balanced, such as, dealing properly with day care situations, not only when a woman gets pregnant, but also when the baby's born, some facilities and even time off...I think if the structures and the system and colleagues and the whole machinery were more conducive to really understanding the needs of women or that family, and make sure that they can spend time with kids, perhaps women would stay.

Danielle, an associate professor at a research-intensive university, regretted postponing starting a family. She talked about her decision, saying,

> I think we sacrificed quite a bit personally. Like right now, I'm pregnant, we're having a baby, it's really exciting, but, sometimes you feel like, I'm 36, right? Why did I wait this long? ...I think we did make a lot of sacrifices personally, and I think we're [at] a stage right now where, granted, we've got to keep everything going and moving forward in the lab, but, I think we just need to take a breath here and assess what's going on... 'cause, you can't get that time back and you have to sit and realize, I think I've gotta physically make time for things other than this [job].

She chose to wait until she was through the tenure process, which despite her regret, did allow her some flexibility in terms of exposure to chemicals. She felt her research group was at a place where it was relatively self-sustaining, allowing her to minimize her time in the lab while pregnant.

Marie, an assistant professor, expressed similar concerns regarding the timing of starting a family and the tenure process.

> ...by the time I get tenure, I'll probably be 36? And I feel, if I waited until I got tenure, and then I got tenure, and then I had kids, then everything would be great. But, if I waited until I got tenure, and then I didn't get tenure, then I would have put off something that was really, really important to me and then I wouldn't have [kids] and I wouldn't have a job, so then I think I would be even more upset.

Interestingly, Marie did not discuss safety issues even though she works with very hazardous materials in her research lab.

Petra echoed the challenges with the tenure system and the university structure with regard to women getting pregnant and starting families. She advocated for more supportive and flexible pathways to tenure, saying,

> I don't think a woman should ever compromise [her] biological clock for her job, but that's why I [support] having better structures that would facilitate all aspects of having children, when you're expecting and after you have them. I think [this is] sorely needed, because,

frankly, I think women are just discouraged from having kids when they are in [a] tenure-track situation. They are afraid about...the repercussions on their future. I had my second child [before tenure] because I wanted to have a second child and I was not young compared to American students, so it was just either you do it now or you won't do it, but if I had been a little younger, perhaps I would have waited. And some people may think, justly, that it's not fair that you have to wait to have kids because of your job. So I totally sympathize with that.

Catherine, a professor and department chair, discussed how her career and the challenge of finding jobs in the same location for her and her husband led to them having a smaller family than planned. She said,

...my husband and I lived apart for a number of years trying to get jobs in the same place, and I would definitely say I had fewer children than I would [liked to] have [had]. I have one daughter, I have fewer children than I would have [had], I think, if I weren't at a research institution...I had one fewer child than I wanted.

Again, the career and institutional structures have changed women's plans for starting and building a family.

While women at research institutions felt there were significant challenges with regard to being a professor and mother, they did not bring up the issues of safety that were highlighted by women at PUIs. While these safety challenges were not nonexistent, they were less salient for professors at research universities. One primary reason for this difference between faculty concerns is the structure of research labs at PUIs compared to research institutions. At PUIs, faculty maintain the research program continuously. They tend to work with inexperienced and transient undergraduates, which means that they are in the research lab training student researchers and maintaining research progress when in between students. At research institutions, postdocs and experienced graduate students can help train new graduate students and undergraduate researchers. This lessens the need for PIs to be physically in the research lab, affording more flexibility during and post-pregnancy.

Additionally, instructional roles are very different between PUIs and research institutions. As noted previously, instructional labs at research universities are usually taught by graduate teaching assistants, particularly at the introductory levels. At PUIs, lab courses may or may not be taught by undergraduate teaching assistants at the introductory level. More likely than not, faculty members need to be present in the instructional laboratory. As a result of this structure, it is again more flexible for faculty at research institutions to avoid contact with hazardous chemicals during pregnancy and breast-feeding. The main concern for women at research institutions was the high research expectations for tenure and deciding to postpone starting a family, whereas for women at PUIs, safety in the presence of hazardous chemicals was the primary concern. If the tenure process at research institutions was seen as more family-friendly and supportive of women, it's likely that safety would become more of a focal point for women at these institutions.

Current Policies and Strategies for Dealing with Safety Challenges

One major challenge for women chemists is the lack of information regarding hazards during pregnancy. Publicly available recommendations are few and far between, and those that are available are vague. They leave the determination of safety up to the individual, but it is a challenging position to be in to decide the fate of an unborn child when you have incomplete or minimal information. For example, Virginia Polytechnic University has a public statement on Safe Pregnancy for Laboratory Workers (http://www.chem.vt.edu/facilities/resources/safety-pregnancy-accommodation.pdf). It is just over a page long and includes the following paragraph:

> The Chemistry Department seeks to minimize the risks of working in its laboratories for all employees and students, especially for pregnant women because of the known sensitivity of the fetus to specific chemicals, in particular teratogens. All laboratory workers are expected to know the hazards of chemicals they are using, including the pregnant woman. Material Safety Data Sheets (MSDS) are essential, but may not provide a complete set of recommendations. Additional protective equipment may be available, but alternatives to laboratory work such as spectroscopic or computational studies, library work, writing, or seminar preparation may be requested by the pregnant laboratory worker. Each woman's situation will be different, so the Department can be creative and flexible. We encourage a pregnant woman to consider those accommodations that she might request for her well being, and for the well being of her fetus.

Also in this statement are recommendations for pregnant graduate students, encouraging them to consult with their research director or graduate program director regarding questions they may have. As has been discussed, research and graduate program directors are likely to have few answers and may not be able to provide answers that graduate students are seeking.

Also available online is a website from The University of California, San Francisco Office of Research: Environmental Health and Safety (http://or.ucsf.edu/ehs/9399-DSY/11389). It provides the following statement regarding pregnant women working with chemicals:

> Pregnant workers should avoid unnecessary exposure to chemicals. Since the beginning of the 20th century, thousands of new synthetic chemicals have been developed, and only a small portion of these chemicals have been adequately studied to determine whether they pose a risk of cancer or birth defects. Therefore, it is advisable to limit any unnecessary chemical exposure during pregnancy. Some chemicals are well known to increase the risk of cancer or birth defects.

While these types of statements are meant to be helpful, the lack of concrete recommendations can be unsettling for women. There are few chemistry departments or universities that have a statement regarding pregnant workers at all, so it is positive that some statements exist to address the needs of pregnant women in chemistry. It is more common to find recommendations for accommodating graduate students rather than faculty members. Currently, the American Chemical Society does not have a specific document regarding reproductive hazards in the chemical workplace, although there is a list of websites and resources related to reproductive health available on the ACS website. The lack of resources is highly problematic for women trying to make decisions about safety during pregnancy and breastfeeding.

Recommendations for Departments and Administrators

As seen here, there are few concrete recommendations for women chemists regarding reproductive health, pregnancy, and breast-feeding. This poses a unique challenge for women chemists, as well as departmental and university administrators who must help find accommodations for them. The hazards posed by many chemicals during pregnancy and breast-feeding are unknown and are likely to remain unknown. It is up to individuals to decide what they are comfortable with in terms of exposure to chemicals during pregnancy and breast-feeding. This is an uncomfortable position for women to be in, as they are trying to make difficult decisions with incomplete data.

From interviews with chemistry professors who are also mothers, it is clear that open communication between faculty members and department chairs is key to addressing safety concerns and making accommodations during pregnancy. It is also imperative to foster an environment of support and community, where women feel they can communicate with administrators and have support from colleagues. There needs to be recognition of the challenges inherent in pregnancy and motherhood across departments and universities so that adequate support systems and accommodations can be implemented.

As seen from faculty members' reports, the challenges to motherhood are partially systemic, due to structural issues with tenure. One of the major challenges is the disruption of research progress, especially during the pre-tenure years. If research progress stops for motherhood, it can be hard to get it restarted. Even stopping the tenure clock does not fix this. Timing is important in the sciences, and breaks from research pose significant challenges in research progress, publishing, and grant awards. This is a very real challenge outside of the safety concerns women chemists face.

Conclusions

In conclusion, pregnancy and breast-feeding pose unique challenges to women chemists. These challenges are exacerbated by incomplete information about chemical hazards, a lack of policies and concrete recommendations, and the tenure system structure. Women have chosen accommodations based on their individual needs and concerns. While it is appropriate to be flexible and meet each individual's circumstances, it is problematic to not have overarching policies or recommendations. It leaves each woman to advocate for herself and be subject to colleagues, chairs, and administrators who may be less than accommodating.

As seen from the reports of women faculty, there are several strategies used to address the safety concerns of working in the laboratory during pregnancy. Finding alternate teaching or service responsibilities and avoiding chemical exposure was a popular choice, but it came at the expense of research progress. At research-intensive institutions, it was more common to postpone starting a family until after tenure based on the rigorous tenure expectations as well as the added flexibility that tenure brings. These women stressed the importance of communication and a supportive chair in finding accommodations they were comfortable with during pregnancy and breast-feeding. Rather than waiting for a faculty member to become pregnant and address it at that time, discussions at the department and university level are encouraged to develop possible strategies. It is recommended that departments include a statement about accommodating pregnancy in their policy statements. This would not only provide clearer strategies for handling pregnancy, but would demonstrate a willingness to support female faculty members, a critical element in recruiting and retaining female chemists. Finally, a statement or policy from ACS or other professional organizations is encouraged as a way to recognize the challenges mothers in chemistry face and to open discussion about how departments can best accommodate and support female faculty.

Main Steps in Megan's Career

Megan is an Assistant Professor at Western Michigan University, with a joint appointment in the Department of Chemistry and the Mallinson Institute for Science Education. She completed her B.S. at the University of Indianapolis, where she was an All-American swimmer and NCAA Woman of the Year finalist. She attended Indiana University School of Medicine for a semester before starting her graduate work at Purdue University. She completed her M.S. and Ph.D. in Chemistry at Purdue University, conducting Chemical Education Research. Before starting her position at WMU, she was a postdoctoral researcher at the American Chemical Society Examinations Institute at Iowa State University. Current projects at WMU include women chemists' academic and career choices, motivational theories in undergraduate and graduate STEM education, feminist critiques of academic science, laboratory curriculum development, graduate student professional development, student outcomes from participating in Problem-Based Learning laboratory units, and instructor and institutional adoption of evidence-based teaching practices.